Watermaths

Watermaths
Process Fundamentals for the Design and Operation of Water and Wastewater Treatment Technologies

Third Edition

Simon Judd
Cranfield Water Science Institute
Cranfield University

Professor Simon Judd has taught at Cranfield University in the UK and at Qatar University in the Arabian Gulf. This book is based on 27 years' experience in teaching the fundamentals of water and wastewater technologies.

Published by

IWA Publishing
Alliance House
12 Caxton Street
London SW1H 0QS, UK
Telephone: +44 (0)20 7654 5500
Fax: +44 (0)20 7654 5555
Email: publications@iwap.co.uk
Web: www.iwapublishing.com

First published 2019
© 2019 IWA Publishing

Disclaimer

The information provided and the opinions given in this publication are not necessarily those of IWA and should not be acted upon without independent consideration and professional advice. IWA and the Editors and Authors will not accept responsibility for any loss or damage suffered by any person acting or refraining from acting upon any material contained in this publication.

British Library Cataloguing in Publication Data
A CIP catalogue record for this book is available from the British Library

ISBN: 9781789060386 (Paperback)
ISBN: 9781789060393 (eBook)
ISBN: 9781789060409 (ePub)

Graphics by Oliver Judd

Cover image: iStockphoto.com

Contents

Chapter 3
Fluid physics **17**

Chapter 4
Chemical stoichiometry and equilibria **29**

Chapter 5
Chemical and biochemical kinetics … **51**

Chapter 8
Reactor theory **77**

Chapter 9
Cost analysis **83**

Chapter 10
Solutions to exercises **89**

List of Tables

List of Figures

List of Abbreviations

AC	Activated carbon
ASP	Activated sludge process
BOD	Biochemical oxygen demand
CAS	Conventional activated sludge
CAPEX	Capital expenditure
COD	Chemical oxygen demand
CSTR	Continuous stirred tank reactor
DO	Dissolved oxygen
DOC	Dissolved organic carbon
EBCT	Empty bed contact time
GAC	Granular activated carbon
HRT	Hydraulic retention time
IFAS	Integrated fixed-film activated sludge
LMH	Litres per m^2 per hour
LHS	Left-hand side (of equation)
LSI	Langelier Saturation Index
MBR	Membrane bioreactor
MBBR	Moving bed bioreactor
MF	Microfiltration
MGD	Megagallons per day
MLD	Megalitres per day
MLSS	Mixed liquor suspended solids
MLVSS	Mixed liquor volatile suspended solids
NF	Nanofiltration
ON	Oxidation number
OPEX	Operating expenditure
PAC	Powdered activated carbon
PFR	Plug-flow reactor
psi	Pounds force per square inch

UF	Ultrafiltration
UV	Ultraviolet
RHS	Right-hand side (of equation)
RO	Reverse osmosis
SCFM	Standard cubic feet per minute
SBR	Sequencing batch reactor
SRT	Solids retention time
TDS	Total dissolved solids
TKN	Total Kjeldahl nitrogen
TN	Total nitrogen
TOC	Total organic carbon itself
TSS	Total suspended solids
TF	Trickling filter

UNITS, SI

The key SI parameters and their units are given below. Non-SI units are discussed in Section 2.4.2. Chemical concentration conversions are depicted in Figure 14.

Length

µm	micrometre	10^{-6} m
Cm	centimetre	10^{-3} m
mm	millimetre	10^{-2} m
km	kilometre	10^{3} m

Force

N	newton	$kg \cdot m/s^2$

Power

W	watt	$kg \cdot m^2/s^3$
		or J/s

Mass

µg	microgram	10^{-6} g
mg	milligram	10^{-3} g
kg	kilogram	10^{3} g
te	tonne	10^{6} g

Pressure

Pa	pascal	$kg/(m \cdot s^2)$
		or N/m^2

Chemical concentration

M	Molarity	mol/L, also denoted "[ ]"
–	Equiv. conc.	meq/L
–	Conductivity	µS/cm

Volume

mL	millilitres	10^{-3} L
ML	megalitres	10^{6} L
m^3	cubic metres	10^{3} L

Energy

J	joule	$kg \cdot m^2/s^2$
kWh		3.6×10^{6} J

Symbols

A list of symbols and their definitions is given below. Symbols representing more than one parameter depending on the context in which they are used are assigned the relevant equation, table or chapter number(s). Units provided are indicative: alternative/ non-SI units (e.g. hours or days for time, grams or tonnes for mass, etc.) may be used depending on the context. Chemical symbols are excluded: these are listed in Tables 13 and 14.

a	orifice cross-sectional area (Equation 3.14), m^2; number of moles of chemical "A" in general equation (Equation 4.23)
A	cross-sectional area, m^2
Ar	Archimedes number
b	slot width (Equation 3.17), m; number of moles of chemical "B" in general equation (Equation 4.23)
B	flow channel width (Equation 3.17), m
c	chemical concentration, moles/L or mg/L; constant (Equation 5.9); number of moles of chemical "C" in general equation (Equation 4.23)
$c*$	equilibrium concentration,
c_B	biomass concentration, kg/m^3
c_{DO}	DO mass concentration, kg/m^3
C	mass concentration in kg per m^3, g per L, etc
C_p	specific heat capacity, $J/(kg \cdot K)$
d	diameter, m; characteristic length (Table 5), m; number of moles of chemical "D" in general equation (Equation 4.23)
D	diffusion coefficient, m^2/s (Chapter 7); discount factor (Chapter 9)
F	force, $kg/(m \cdot s^2)$
Fr	Froude number
g	acceleration due to gravity (9.81 m/s^2)
h	depth or height of water, m
H	head loss, m (Chapter 3); Henry's Law constant (Equation 4.38)
H_f	frictional head loss along a pipe, m
J, J_1, J_2	constants in hydraulic equations (Chapter 3)
J_1, J_2	coefficients in Equations 3.19
k	mass transfer coefficient (Table 5, Chapter 7), m/s; rate constant (Chapter 5, Chapter 8), s^{-1}.
$k_{fittings}$	head loss coefficient for fittings
k_h	head loss coefficient for filter media
$k_L a$	gas absorption coefficient, s^{-1}

k_T	thermal conductivity, $W/(m \cdot K)$
K	coefficient (Equation 3.21); equilibrium constant (Equations 4.24, 4.49); overall mass transfer coefficient (Chapter 7)
K_a	acid dissociation constant
K_b	base dissociation constant
K_F	Freundlich constant, $(mg/g)(L/mg)^{1/n}$
K_G	total gas transfer coefficient
K_L	total liquid transfer coefficient
K_s	Half saturation coefficient
K_{SP}	solubility product
l, L	length, m
m	mass, kg; coefficient in general power equation (Equation 2.1)
M	mass flow rate, kg/h
M_b	biomass growth rate in the reactor, kg/h
$M\ alk$	M alkalinity
n	number or reactors or membrane elements in series; exponent in general power equation (Equation 2.1)
N	number of weir ends (Equation 3.21); gas transfer rate per unit volume (Equation 5.14), $kg/(m^3 \cdot h)$
p, P	pressure, $kg/(m \cdot s^2)$
P_{atm}	atmospheric pressure, $kg/(m \cdot s^2)$
P_h	hydrostatic pressure, $kg/(m \cdot s^2)$
Pe	Peclet number
$P\ alk$	P alkalinity
pH	negative log of the molar hydrogen ion concentration
pK_a	negative log of the acid dissociation constant
pOH	negative log of the molar hydroxide ion concentration
Pr	Prandtl number
q_e	equilibrium adsorbent capacity, mg/g
q_0	oxygen mass flow per unit biomass mass, $kg\ O_2/(kg\ cells \cdot h)$
Q	volume flow rate, m^3/h
r	reaction rate, $mg/(L \cdot h)$
R	regeneration ratio
Re	Reynolds number
s	hydraulic gradient (Chapter 3), m/m, or substrate concentration (Chapters 5 & 6)
s_i, s_R	concentration of limiting nutrient (or substrate) in influent, reactor, $kg \cdot m^{-3}$
S	Project cost, \$
Sc	Schmidt number
Sh	Sherwood number
t	time, s, min, h or d
t_d	population doubling time, min, h or d
T	temperature, °C or K
v	velocity, m/s or m/h
v_a	approach velocity, m/s or m/h
V	volume, m^3
w	width, m
W	resin mass, kg
We	Weber number
x, x_t	cell population, cell population at time t, mL^{-1}

x_f	carrying capacity, mL^{-1}
$X (X')$	Mixed liquor (volatile) suspended solids
y	non-ideality coefficient for conductivity
Y	yield coefficient; year in Equation 9
α	(Membrane) conversion
ε	filter bed porosity
θ, θ_w	residence or retention time, hydraulic retention time, h
θ_s	solids retention time, d
κ	pipe roughness, m
λ	Darcy–Weisbach friction factor
μ	dynamic viscosity, $kg/(m \cdot s)$; specific growth rate, h^{-1} (Chapter 5)
μ_m	maximum specific growth rate, h^{-1}
υ	kinematic viscosity, m^2/s
ρ	density, kg/m^3
ρ_s	solids or particulate density, kg/m^3
σ	surface tension, kg/s^2; media permeability (Chapter 3), m/s
τ	shear stress, $kg/(m \cdot s^2)$

Preface

A number of books on the market are dedicated to water or wastewater treatment. The benchmark for potable treatment is arguably the MWH book *Water Treatment, Principles and Design* (3rd edn, 2012, Wiley), by Crittenden, Trussell, Hand, Howe and Tchobanoglous, whilst for wastewater it is unquestionably the Metcalf and Eddy tome *Wastewater Engineering: Treatment and Resource Recovery* by Burton, Stensel, Tchobanoglous and Tsuchihashi (2014, McGraw-Hill).

There are several other books addressing the mathematics of water and wastewater treatment plant specifically. These include Shun Dar Lin's *Water and Wastewater Calculations Manual* (3rd edn, 2014, McGraw-Hill), and Frank Spellman's *Mathematics Manual for Water and Wastewater Treatment Plant Operators: Water Treatment Operations: Math Concepts and Calculations* (2014, CRC Press). These books are both comprehensive, informative and tend to be targeted at practitioners or at least assume some knowledge of the industry. Books targeted directly at plant operators include Bob Larsen's *Math Handbook for Water System Operators* (Outskirts Press, 2010) and Jerry Grant's engaging *Lumpy Water Math* (Jerry Grant, 2007).

On the face of it, then, there is perhaps little room for another textbook in this area. However, *Watermaths* is not designed to be a comprehensive handbook, and assumes no knowledge of the industry. The process technologies are not described in any detail, since many other textbooks – including some of those identified above – already do this. Instead, this book aims to provide readers with enough information to enable them to tackle the basic calculations used in the design and operation of water/wastewater unit operations. It is thus a primer for all those interested in technology design and operation in the water industry.

Little or no prior knowledge of science or engineering is required, and a nomenclature is provided for all symbols used in the text. The book is restricted to SI units, rather than US/Imperial units as many of the above books are, and divided into nine chapters based on the core subjects or disciplines in this area, rather than according to either the unit operations or their applications. Within each chapter are a number of example calculations followed by exercises, intended to reinforce the learning, for which solutions are given in the final chapter. Exercises range from simple single-calculational step problems to more complex ones, and there is an over-arching design problem based on an advanced water reuse process to provide context. The calculations form a large part of the book, and are supplemented by a few illustrations and sufficient background to explain the purpose of the calculations and ultimately tackle the design problem.

It is hoped that this book, the third edition, delivers the skills required with a light touch and a simple and succinct delivery that gets the reader up to speed as rapidly as possible. The book is targeted at undergraduates, graduates and practitioners with an interest in water pollution control and freshwater supply technologies, but possibly lacking a physical sciences or engineering knowledge base. It is strictly limited to the most essential areas of knowledge required for process technology design and operation, containing tuition in basic numeracy, as well as chemistry, process engineering and fluid physics. Lastly, the book should be accessible and, hopefully, not break the bank for anyone looking to buy it.

© IWA Publishing 2019. Watermaths: Process Fundamentals for the Design and Operation of Water and Wastewater Treatment Technologies
Author: Simon Judd
doi: 10.2166/9781789060393_xxiii

Chapter 1

Introduction

1.1 WATER AND WASTEWATER TREATMENT PROCESSES

The treatment of water or wastewater relies on a number of individual unit operations which are combined to make a process, often referred to as a *process treatment scheme*. Although seemingly complicated and diverse in practice, these unit operations are all work according to a relatively narrow range of governing principles – whether water from the ground, lakes, reservoirs, rivers or the sea is to be purified for drinking, or wastewater (i.e. sewage or industrial effluent) is to be cleansed for safe discharge to the environment. These principles, or disciplines, can be summarised as physical separation (primarily *sedimentation* and *filtration*), fluid physics (*hydraulics, rheology*), chemistry or biochemistry (*stoichiometry, kinetics, equilibrium thermodynamics*), and chemical engineering (*reactors, mass balance, mass transfer*). Whilst this list may not be comprehensive, most common water and wastewater treatment processes are based on these core areas of science and engineering. It is also the case that the design of the unit operations based on such principles demands only a cursory knowledge of mathematics – in essence, just arithmetic (i.e. calculations) and the manipulation of equations.

Fluid physics (Chapter 3) and mass balance (Chapter 6) define the transport of liquid and solids in and between unit operations. The basis of the unit operations, and specifically the way in which contaminants are removed from the water, is defined almost entirely by physical separation and chemistry/biochemistry (Figure 1.1). Thus, suspended solids in water are removed either by settling (sedimentation or gravitation, Section 7.5) or filtration, often through a bed of sand (i.e. depth filtration, Section 3.7). Dissolved materials may also be removed by adsorption (Section 4.6.8), and in this case the action of the adsorbent (i.e. the medium adsorbing the pollutant) is based on physical chemistry.

Pollutants may also be transformed into either less onerous products or ones which can be removed physically by a subsequent unit operation. The transformation of a dissolved chemical into another (sometimes mis-named "destruction") is a key part of chemistry, and the chemical action of an organism affecting such a transformation is the basis of biochemistry. In general, these processes tend to oxidise (i.e. add oxygen to) the pollutant, and biochemical reactions in particular can completely oxidise organic (i.e. large carbon-based molecular) pollutants to the most simple *mineral* (inorganic) end products. Many chemical oxidants, such as chlorine and ozone, have the power to inactivate (i.e. kill) bacteria and viruses and are thus defined as *disinfectants*. Disinfection may also be achieved by UV irradiation. Finally, some chemical reactions do not involve oxidation or reduction, but simply adjust the chemical conditions of the water, such as its acidity or the solubility of the pollutants.

Chemical and biochemical reactions are represented by equations, analogous to mathematical equations, which define the ratio of a chemical reagent used to treat the water with the chemical pollutant it is transforming. This is referred to as *stoichiometry* (Section 4.5). Other aspects of chemistry are the rate at which reactions proceed, referred to as *kinetics* (Chapter 5), and the balance of reactants and products at the end of the reaction, referred to as *equilibrium thermodynamics* (Section 4.6).

© IWA Publishing 2019. Watermaths: Process Fundamentals for the Design and Operation of Water and Wastewater Treatment Technologies
Author: Simon Judd
doi: 10.2166/9781789060393_0001

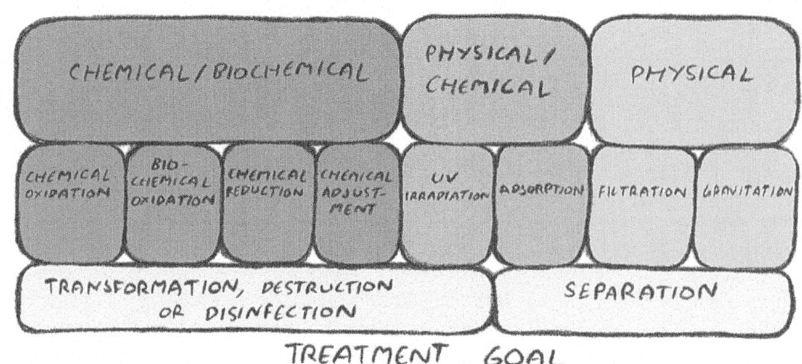

Figure 1.1 Unit operation principles in water and wastewater treatment.

Whilst all of these subjects, if explored comprehensively, are very broad and encompass a wide variety of topics, when applied to water and wastewater treatment process technology design and operation the scope is quite limited (Table 1.1). The following chapters provide sufficient information to enable the basic design of water and wastewater treatment processes based on the most common unit operations. By way of demonstrating this, the design and operational elements of a complete installation for advanced wastewater treatment for water recovery and reuse will be considered as an example in a subsequent section (Section 1.4).

It is the governing principles (Table 1.1) by which a unit operation acts to remove or transform pollutants that determine the way in which the process is designed and/or operated. As indicated in Figure 1.1, they are almost all either physical or (bio) chemical in nature.

Gravitational settling represents the simplest of all processes, requiring only a large tank and a means of removing both the *clarified* (or *supernatant*) product water and the settled solids (or sludge). The product water normally flows over a weir into a channel, and the solids scraped from the bottom of the tank into a trough. The tanks can be either cylindrical (Figure 1.2a) or rectangular (Figure 1.2b) in shape, and in both cases the residence time of the water flowing through the tank must be long enough to allow bulk of the solids to gravitate to the tank base.

Filtration is either by sieving (surface filtration) or adsorption throughout the depth of a media bed. Examples of unit operation technologies based on surface filtration include hollow fibre membranes (Figure 1.3a), used for filtering water,

Table 1.1 Classical unit operations and their main governing disciplines/subjects.

Unit Operation	Fluid physics	Physical separation[1]	(Bio) chemistry[2]	Chemical engineering[3]
Pumping	X			
Screening	X	SF		
Sedimentation/flotation	X	Gr		MB
Scrubbing/stripping	X		St	MB, MT
Biological treatment			St, K	MB, MT, R
Coagulation/flocculation			St, ET, K	R, MB
Acidity adjustment			ET	
Activated carbon	X	DF	ET	
Sand filtration	X	DF		
Membrane treatment	X	SF	ET[4]	MT[4], MB
Chemical disinfection			St, K	
UV irradiation			K	MT
Sludge thickening/dewatering	X	SF, Gr[5]		MB
Sludge digestion			St, K	MB, MT, R

[1]SF, surface filtration (sieving); DF, depth filtration (adsorptive filtration); Gr, gravitation.
[2]St, stoichiometry; K, kinetics; ET, equilibrium thermodynamics.
[3]R, reactors; MB, mass balance; MT, mass transfer.
[4]For reverse osmosis and electrodialysis.
[5]For centrifugation.

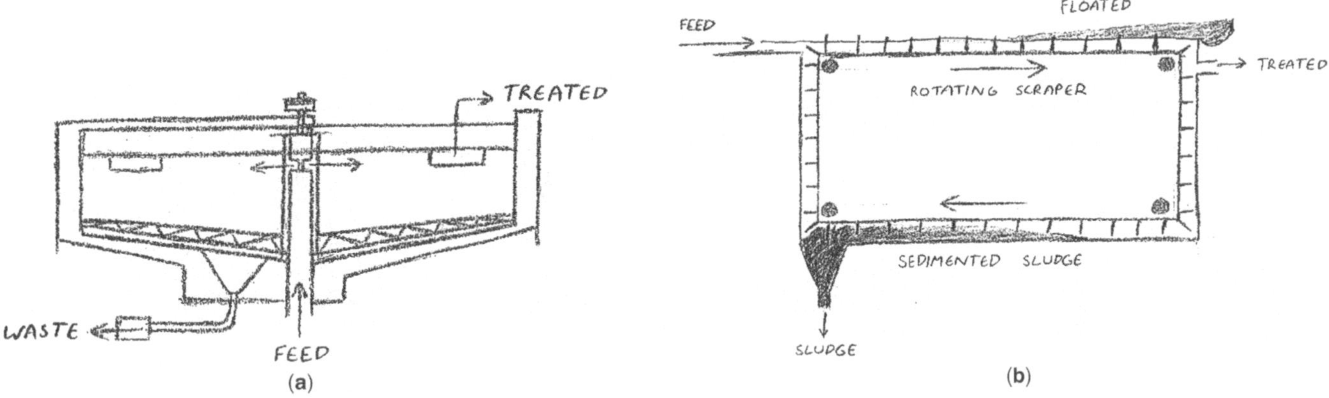

Figure 1.2 Sedimentation tanks: (a) cylindrical; and (b) rectangular.

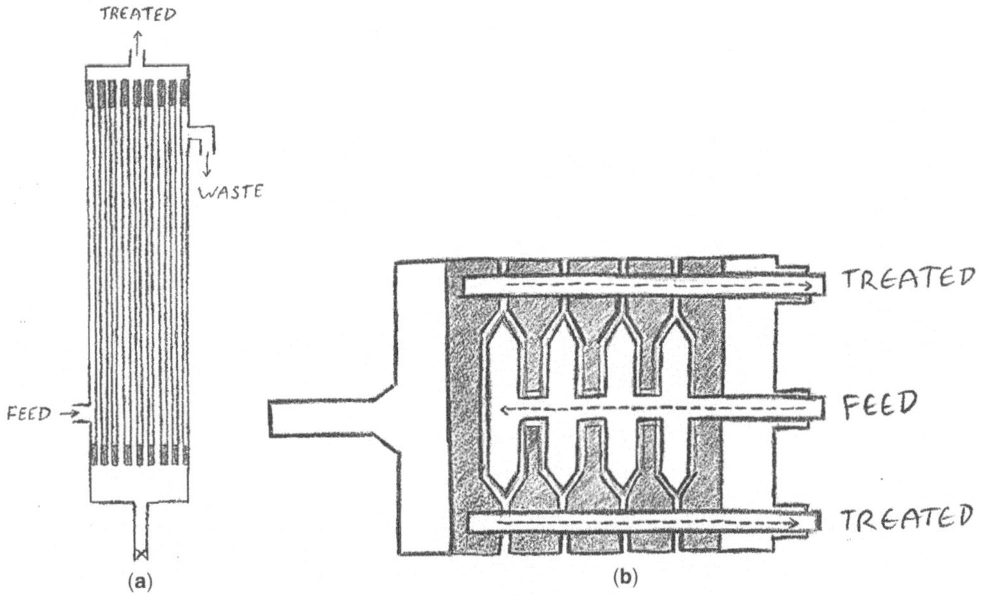

Figure 1.3 Surface filtration technologies: (a) hollow fibre membrane module, and (b) filter press.

screens for the preliminary treatment of sewage, and filter presses, used for filtering (or dewatering) sludge (Figure 1.3b). For all these processes the rejected solids form a layer on the surface of the filter and need to be periodically removed. In the case of depth (or media) filtration the particles are removed throughout the depth of the media (often sand). Particle removal increases with increasing media depth, the depth being 1.5–2.5 metres in the case of sand filters. As with sedimentation tanks, the media vessels can be either rectangular (Figure 1.4a) or cylindrical (Figure 1.4b).

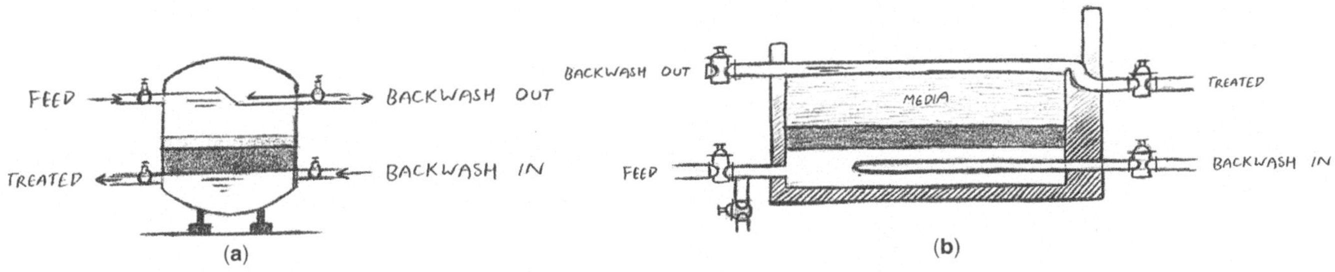

Figure 1.4 Depth filtration vessels: (a) cylindrical; and (b) rectangular media filters.

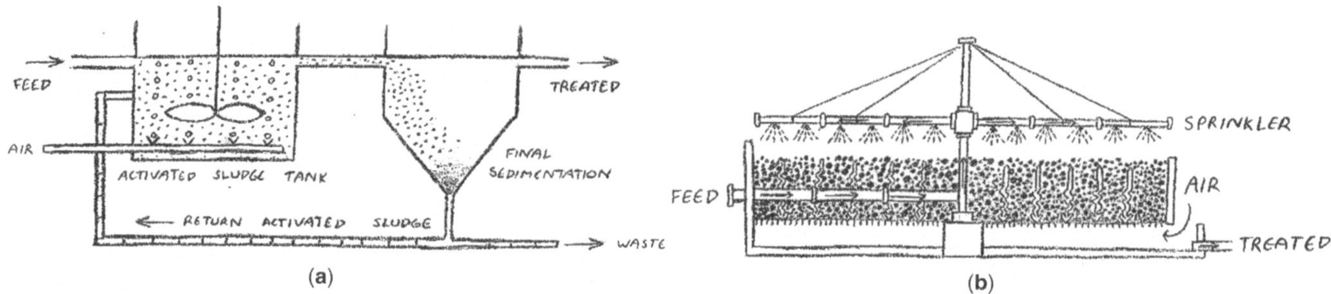

Figure 1.5 Biochemical oxidation technologies: (a) *suspended growth* (the activated sludge process); and (b) *fixed film* (the trickling filter).

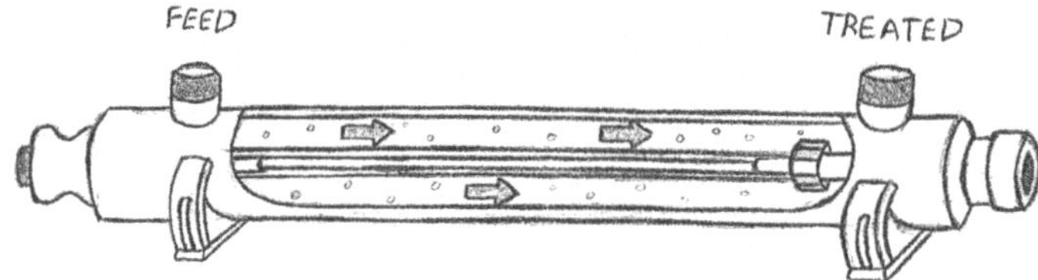

Figure 1.6 UV irradiation chamber.

Adsorption processes can be configured either as media beds – analogous to sand filters – or as a powder or slurry of material dosed into a mixer tank. The powder is subsequently removed by filtration. Dissolved organic and ionic materials can both be removed by adsorption. If the adsorption process is configured as a media bed then, as with media filters, removal takes place throughout the media depth. Media for organic carbon removal are normally grains of activated carbon, hence *granular activated carbon* (GAC). For ion removal the media are often *ion exchange* beads (which are completely spherical).

Biochemical oxidation is the most fundamentally important process in the treatment of wastewater. In this process active micro-organisms are retained dispersed particles (or *flocs*) in a tank (a *suspended growth* process), as in the widely employed *activated sludge process* (Figure 1.5a), or formed on the surface of some media (as a *fixed film*), for example in a *trickling filter* (Figure 1.5b). The micro-organisms oxidise organic carbon to inorganic carbon (carbon dioxide and bicarbonate) and ammonia to nitrate. The conversion rates for these processes are slower than for most chemical processes, such that the residence time in the tank is several hours.

Chemical oxidation employs a reagent – which can be a powerful oxidant such as ozone, chlorine or hydrogen peroxide or a milder oxidant such as oxygen from air – to carry out a chemical reaction. If the reagent is gaseous then it is introduced into the reaction vessel or tank as very small bubbles via a *fine bubble diffuser*. If liquid it is delivered via a dosing pump. In either case the reagent concentration and flow rate are controlled to meet the *demand* of the pollutant that is being removed or converted by the oxidation process.

Finally, UV irradiation is unusual in that it is a physical process (the application of light) which effects a change in the biological state of micro-organisms in the water – from active (alive) to inactive (dead), i.e. *disinfection*. The process uses short-wavelength ultraviolet light to destroy the nucleic acids and disrupt the DNA of the micro-organisms, leaving them unable to perform vital cellular functions and preventing them from reproducing. The UV radiation is delivered via a tube, or more normally an array of tubes, held within a cylindrical chamber or rectangular conduit through which the water flows (Figure 1.6).

1.2 UNIT OPERATIONS SYMBOLS

Water and wastewater treatment schemes normally contain a number of different unit operations placed in a sequence to provide purification of the feedwater. To present this diagrammatically requires a recognised set of symbols representing the individual unit operations listed in Table 1.1. These processes may each have a number of different configurations and designs; the symbol can either be a generic symbol for a unit operation or piece of equipment, or it may refer to a specific type. For example, the generic symbol for a pump is normally as shown in Figure 1.7(a). However, there are many different types of pump – including centrifugal, gear, progressive cavity and screw – which can all be depicted specifically (Figs 1.7b–e). Pumping of air is achieved using *blowers*, for low-pressure pumping, and *compressors* for high pressures.

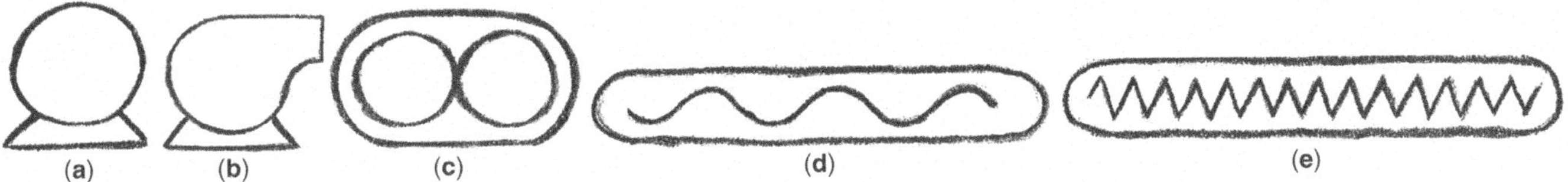

Figure 1.7 Pump symbols: (a) generic symbol, and symbols for (b) centrifugal, (c) gear, (d) progressive cavity, and (e) screw pump.

The symbols appear in diagrammatic representations of water and wastewater treatment schemes, either in a *process flow diagram* (PFD) or a *process and instrumentation diagram* (P&ID). A PFD shows the unit operations and the flow of materials (water, air, sludge and chemical reagents) between them, as well as the pumps and often fittings such as valves and regulators. A P&ID includes further details of the equipment specification, including the pump type, pipework diameter and discharge routes (drains and vents), as well as the control and monitoring instruments, such as pressure and temperature sensors.

P&IDs are required to provide comprehensive design information for implementing a process, and use largely consistent symbols and notation for all the equipment involved. PFDs, on the other hand, can use either recognisable symbols or simple blocks to represent the individual unit operations, and the symbols used are not necessarily consistent across all users or design software suppliers. Symbols used for the unit operations relating to this book are shown in Figure 1.8.

1.3 WATER QUALITY

Given that water and wastewater treatment processes are ultimately designed to remove contaminants from water, the nature of these and how they may be removed must be considered. Contaminants may be defined in many ways, but in general are either dissolved or suspended. If suspended then they may be removed by solid–liquid separation processes (sedimentation, filtration or *centrifugation* – an enhanced gravitation process). If dissolved then there are a number of options:

(a) Conversion to suspended materials, generally using chemicals to precipitate the contaminant as a solid or gas,
(b) Adsorption onto a solid material, which may either be suspended or fixed as a bed (such as granular activated carbon or sand), or
(c) Conversion to relatively innocuous end products, such as carbon dioxide, nitrogen and water.

The concentration of suspended materials in water is usually defined as the *total suspended solids*, or TSS. The equivalent term for dissolved solids is thus TDS. However, if dissolved materials are organic in origin then their concentrations may be defined either as the concentration of organic carbon itself (TOC for *total organic carbon*, DOC for *dissolved organic carbon*), or by the oxygen this carbon uses up when it is oxidised – the ultimate end oxidation product being carbon dioxide (CO_2). For the latter, the concentration is defined by the *oxygen demand* of the pollutant in mg of oxygen/L. This can be measured either as the *chemical oxygen demand* (COD), if oxidation uses relatively aggressive chemical conditions, or the *biochemical oxygen demand* (BOD) if oxidation is based on the action of micro-organisms using oxygen to biodegrade the pollutant. The BOD is a key measurement in biological treatment processes where removal of organic carbon relies on micro-organisms. Similarly, the concentration of biologically degradable nitrogen is expressed as the *total Kjeldahl nitrogen* (TKN). This relates to the amount of oxygen needed to biodegrade certain nitrogen-containing organic compounds (specifically *amino* compounds) to nitrate. In doing so, the TKN contributes to the BOD.

Concentrations of other dissolved components are often expressed with reference to the actual contaminants, normally in mg/L. A key exception to this is the hydrogen ion concentration, which represents acidity and which is always expressed logarithmically (Section 2.3) as the *pH* (Section 4.6.3). Other key inorganic chemical water properties may represent the sum of a number of related species. For example, *alkalinity* (Section 4.6.4) represents the total concentration of species capable of quenching acid. In practice, this normally means bicarbonate (HCO_3^-), carbonate (CO_3^{2-}) and hydroxide (OH^-). Alkalinity has considerable significance in water and wastewater treatment, since it helps to maintain the pH at *neutral* levels (i.e. pH $\sim$7, see Section 4.5.2.1).

1.4 ASSIGNMENT: WASTEWATER REUSE PLANT

The key aspects of an installation to be used for the reuse of municipal wastewater are given below. Questions on specific design and operation parameters follow.

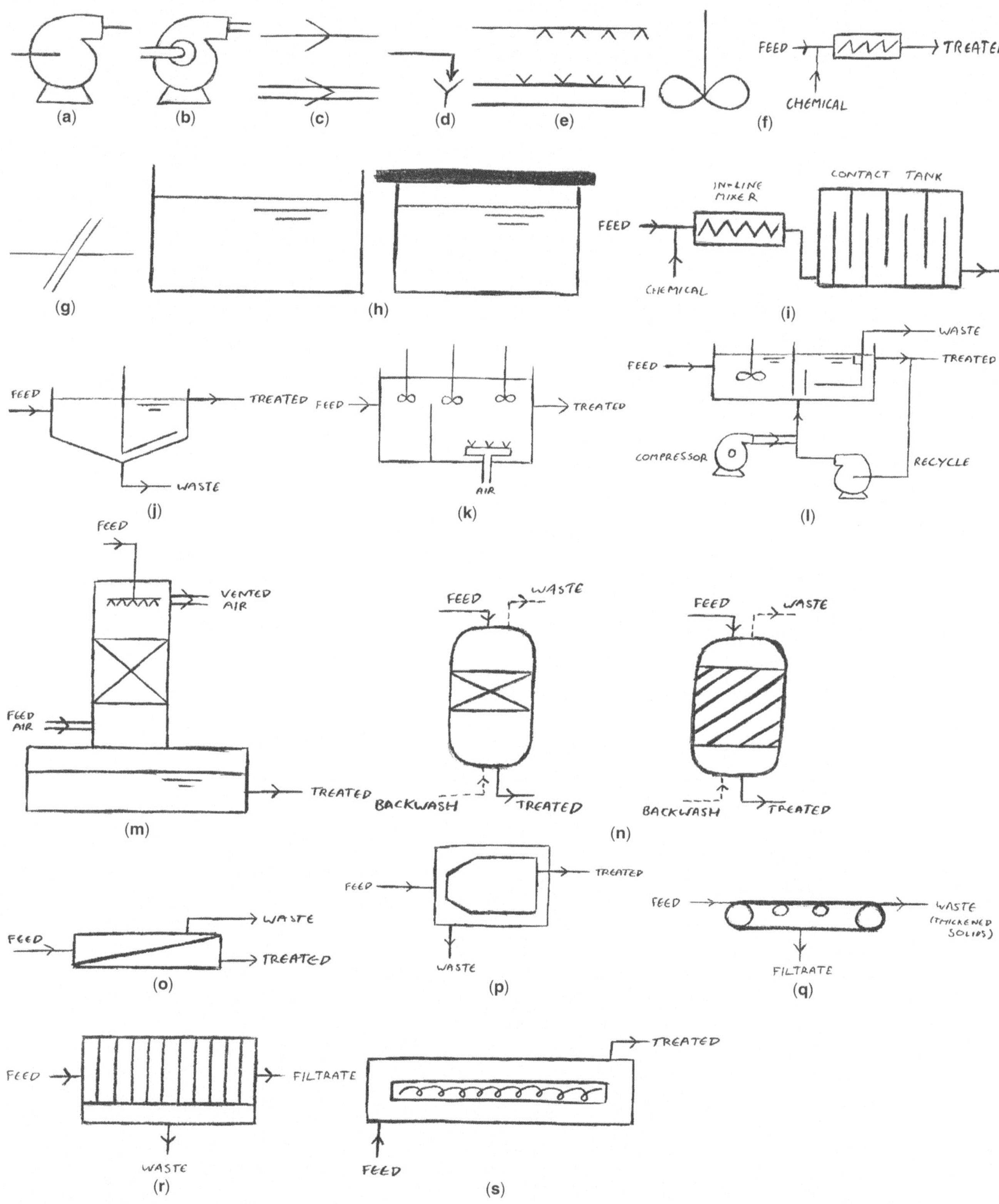

Figure 1.8 Examples of symbols for unit operations used in process flow diagrams: (a) liquid pump, (b) air/gas pump (or blower), (c) pipework (for liquid and gases), (d) discharge point, (e) spray and aerator, (f) mixer (rotating impeller and static in-line), (g) screen, (h) storage tank (open and closed), (i) mixer and contact tank (for dosing chemicals), (j) sedimentation tank, (k) biological process tank, (l) dissolved air flotation tank, (m) gas stripper, (n) media filter (two different depictions), (o) membrane separation, (p) centrifuge, (q) belt thickener, (r) filter press, (s) UV irradiation tube.

Plant features

Wastewater at 12°C containing 240 mg/L BOD, 0.5 mg/L dissolved oxygen (DO), 220 mg/L of alkalinity "as CaCO₃" and having a pH of 7.3 is being pumped from a holding tank to a small treatment works 200 m away. It is transported at a rate of 500 m³/h via a 300 mm diameter galvanised steel pipe fitted with ten long-radius 90-degree bends and four close-radius 45 degree bends, with a sudden enlargement (1:5 ratio) at the pipe exit. It exits into a 5 m deep rectangular biological process tank, with an aspect ratio (the ratio of the length to the width) of 2:1, providing a residence time of 9 hours.

The biological treatment process is fitted with air diffusers delivering an oxygen mass transfer coefficient of 0.00073/s. 1.75% of the feedwater volume forms the sludge waste stream, which is 20% higher in concentration than the biological process tank MLSS concentration of 3.5 g/L. Following biological treatment the residual total suspended solids is 20 mg/L.

The biologically treated water is dosed with 20 mg/L of aluminium sulphate "as Al" at a pH of 6. It then passes through 20 cylindrical sand filter beds, of 0.6:1 length:diameter aspect ratio, operating at a pressure drop of 0.2 bar for the clean filter, the mean sand grain size being 650 μm and the bed porosity 30%.

The water is then chlorinated to provide a 4-log removal of a target micro-organism having a half-life of 180 s. The residual hypochlorite is completely quenched by dosing with 10% stoichiometric excess of sodium bisulphite (NaHSO₃), which is converted to sodium sulphate in the process while chlorine is reduced to chloride.

The water is then pumped to a reverse osmosis array fitted with membrane elements having a rated feed flow capacity of 3.8 L/s and providing a mean conversion of 8% per element. The minimum required treated flow rate from the array is 350 m³/h.

Finally, the treated RO water passes through a degasser to strip the CO₂ using air containing 0.25% CO₂ flowing at 23 kg/min.

Questions

1. *What is the head loss along the length of the pipe?*
2. *What are the dimensions of the process biological tank? If it behaves as a CSTR (continuous stirred tank reactor) and the reaction rate constant for BOD removal is 0.03 per minute, how much BOD remains after treatment?*
3. *What is the aeration rate if the mean dissolved oxygen concentration is kept at 1.8 mg/L, the density of air is 1.22 kg/m³, and oxygen transfer is 55% efficient? It can be assumed that air contains ~23% oxygen by weight.*
4. *What is the solids retention time and the food:micro-organism (F:M) ratio in the bioreactor if the MLVSS/MLSS ratio is 0.75?*
5. *How much acid or alkali is needed for the coagulation step?*
6. *What are the dimensions of the sand filters?*
7. *What is the total volume of the chlorine contact tank?*
8. *How much sodium bisulphite is required if the chlorine residual is around 1.5 mg/L?*
9. *If no more than five RO elements can be fitted per module, what is the most efficient RO array design to achieve the target treated water flow rate?*
10. *If the stripper off-gas contains 5% CO₂ and the feedwater to it has a pH below 4.5, what is the dissolved CO₂ concentration in the product stripped water?*

This is not a typical water or wastewater treatment plant, and includes many apparently significantly different unit operations. However, there only a few disciplines involved in the design of the unit operations and, correspondingly, a limited number of mathematical expressions requiring only a rudimentary command of arithmetic (Chapter 2).

In the first instance, the water has to be transported from, in this case, a holding tank to the treatment plant. Pt 1 therefore demands a knowledge of fluid physics or hydraulics (Chapter 3), which describes the relationship between flow, pressure and energy.

After passing through a tank the water flows into a biotreatment process. Biotreatment uses oxygen from air to carry out biochemical reactions and so demands a knowledge both of biochemistry and, for Pt 3, mass transfer (Chapter 7), since it is mass transfer which determines how much oxygen is available for the biochemical reactions. Also, because the biotreatment process takes place in a reactor, a knowledge of reactor theory (Chapter 8) is needed for Pt 2. Finally, the key biotreatment process design parameters for Pt 4 derive from the flow and fate of materials in the process, which is based on the principle of mass balance (Section 6.2.4).

The effect of dosing with coagulant chemicals (Pt 5) is defined by the chemical reaction (Section 4.5) which provides the ratio of the chemicals involved in the reaction (or stoichiometry). The same principles apply to calculating the amount of quenching agent (Pt 8). The dimensions of the sand filter (Pt 6) are determined by media bed hydraulics (Section 3.7).

Table 1.2 Structure of this book, *chapter number.*

	Mathematics		
	Manipulating equations, *C2*		
	Logarithms & exponents, *C2*		
	Units, *C2*		
Physics	**Chemistry**	**Biology**	**Chemical engineering**
Fluid physics and	Stoichiometry, *C4*	Biokinetics, *C5*	Mass balance, *C6*
hydraulics, *C3*	Equilibrium thermodynamics, *C4*		Mass transfer, *C7*
	Kinetics, *C5*		Reactors, *C8*
	Economics		
	Cost analysis, *C9*		

Chemistry and biochemistry encompass the subject of kinetics (Chapter 5), or the rate at which reactions take place – including the inactivation of micro-organisms (i.e. disinfection); the inactivation rate then defines the disinfection tank size (Pt 7). Both the configuration of the membrane array (Pt 9) and the residual CO_2 concentration (Pt 10) are determined by mass balance under steady-state conditions (Section 6.2.1).

By working through the chapters (Table 1.2) it is hoped that, by the end of Chapter 9, the reader will have sufficient knowledge and competence to complete Questions 1–10 on the process design and operation of the plant. A number of examples are provided within each chapter, along with exercises which range from a single-calculation step to more complex problems. Solutions to the exercises are given at the end of the book, along with the answers to and hints for completing the design assignment.

Chapter 2

Mathematics

For the most part the mathematics associated with this book is purely arithmetic, i.e. the manipulation of numbers. There are four of these areas demanding a level of understanding in water and wastewater treatment technology design and operation:

- Rearranging equations
- Basic geometry
- Manipulation of exponents and logarithms
- Units.

2.1 REARRANGING EQUATIONS

A design problem will often entail the rearrangement of an equation to calculate an unknown parameter. The basic rule of algebraic manipulation is simple: whatever is carried out on one side of the equation must also be conducted on the other side to isolate the parameter of interest. A few common manipulations are given below.

For:	$A = C + B,$	subtract C from both sides:	$B = A - C$
	$A = CB,$	divide both sides by C:	$B = A/C$
	$A = C/B,$	multiply both sides by B/A:	$B = C/A$
	$A = B^n,$	take the reciprocal index:	$B = A^{1/n}$
	$A = \log_{10} B$	raise 10 to the power of A:	$B = 10^A.$

Logarithms are covered in more detail in the following section (Section 2.3), and there are many other types of algebraic manipulations. However, the ones listed above are probably sufficient for most calculations relating to water and wastewater technology design and operation.

© IWA Publishing 2019. Watermaths: Process Fundamentals for the Design and Operation of Water and Wastewater Treatment Technologies
Author: Simon Judd
doi: 10.2166/9781789060393_0009

EXAMPLE: STOKES LAW

This law (see Section 7.5) defines the relationship between the particle diameter d and its settling velocity v_s in water:

$$v_s = \frac{g(\rho_s - \rho_w)d^2}{18\mu}$$

where g, ρ_s, ρ_w and μ can be regarded as constants. If the particle settling velocity is known, then the above equation must be rearranged as follows to find the diameter:

Multiplying both sides by 18μ: $g(\rho_s - \rho_w)d^2 = 18\mu v_s$

Dividing both sides by $g(\rho_s - \rho_w)$: $d^2 = \dfrac{18\mu v_s}{g(\rho_s - \rho_w)}$

Taking the **reciprocal index** (the square root, in this case) **of both sides**: $d = \sqrt{\dfrac{18\mu v_s}{g(\rho_s - \rho_w)}}$ or $\left(\dfrac{18\mu v_s}{g(\rho_s - \rho_w)}\right)^{1/2}$

EXERCISE 2.1

The Swamee–Jain equation defines the characteristics of flow through a channel (Equation 3.11). One form of the equation defines the friction factor λ, a dimensionless number representing the resistance to flow due to the pipework surface roughness, as being related to the pipe diameter d and the Reynolds number Re by:

$$\lambda = \frac{0.25}{\left(\log\left[\dfrac{\kappa}{3.7d} + \dfrac{5.74}{Re^{0.9}}\right]\right)^2}$$

where κ is a constant. How do (a) the diameter, and (b) the Reynolds number depend upon the friction factor?

Note on notation of mathematical formulae

Equations are normally written on the basis of "discrete terms" (exactly as they would be calculated in a spreadsheet package) when presented in linear format. Unless appearing in brackets with other terms, the mathematical functions apply to individual parameters in isolation from neighbouring terms in the equation:

e.g. $5.74/Re^{0.9}$ is $5.74/(Re)^{0.9}$, ***not*** $(5.74/Re)^{0.9}$

        $k/(3.7d) + 5.74/Re^{0.9}$ is $(k/(3.7d)) + (5.74/Re^{0.9})$ ***not*** $(k/3.7d + 5.74)/Re^{0.9}$

        $g\Delta\rho d^2/(18\mu)$ is $[g\Delta\rho(d)^2]/(18\mu)$.

EXERCISE 2.2

The ratio of convective to diffusive transport of a dissolved gas in water flowing through a pipe is given by the Sherwood number (Sh), which is a function of the Reynolds number (Re) and the Schmidt number (Sc):

$$Sh = 0.026\, Re^{0.8}\, Sc^{0.3}$$

Re and Sc are given by:

$$Re = \frac{\rho U d}{\mu} \qquad Sc = \frac{\mu}{\rho D}$$

where ρ and μ are the water density and viscosity, U its flow rate, d the pipe diameter and D the gas diffusion coefficient.
     Calculate the velocity of water flowing through a 0.1 m diameter pipe if it generates a Sherwood number of 4,100. It is assumed that the diffusion coefficient D is $1.9 \times 10^{-9}\ m^2/s$ and the density and viscosity of water are $1,000\ kg/m^3$ and $0.001\ kg/(ms)$ respectively.

Table 2.1 Geometric dimensional relationships ($l=$ length, $w=$ width, $h=$ height).

Shape	Volume	X-sectional area	Surface area[a]	Volume/ surface area
Rectangular prism	lwh	wh	$2wl + 2hl$[b]	$0.5/(h^{-1} + w^{-1})$[c]
Triangular prism	$lwh/2$	$wh/2$	$3hw$	$2l/3$
Cylinder	$\pi d^2 l/4$	$\pi d^2/4$	πdl	$d/4$
Cone	$\pi d^2 l/12$	—	$\pi dl/2$	$d/6$
Sphere	$\pi d^3/6$	—	πd^2	$d/6$

[a]excluding base and any top surface.
[b]$4lw$ when $h = w$.
[c]$h/4$ when $h = w$.

2.2 GEOMETRY

Geometry is essentially the study of shapes. Most shapes in water and wastewater treatment technology are cylindrical (pipes and tanks), rectangular (tanks), occasionally triangular or cone-shaped for some channels and tanks or else, in the case of particles, treated as spherical. Therefore, the key relationships are those for the perimeter, cross-sectional area and volume of these shapes (Table 2.1).

EXAMPLE: VOLUME OF A TANK

A 7 m diameter sedimentation tank comprises a 3.5 m high cylindrical vessel with a 2.5 m high inverted conical base. What is the overall tank volume?

According to Table 2.1:

Cylindrical part: $\pi \times 7^2 \times 3.5/4 = 135$ m^3
Conical part: $\pi \times 7^2 \times 2.5/12 = 32.1$ m^3
So, total is: $135 + 32.1 = 167.1$ m^3

EXERCISE 2.3
A 6 m high open cylindrical tank with walls 75 mm thick must be fitted into an 8 m by 10 m floor space. What is the volume of the largest cylindrical tank that can be fitted into this space? How big a volume would a 6 m high rectangular tank of the same height and wall thickness provide if it completely occupied the space available?

2.3 EXPONENTS AND LOGARITHMS

Exponents and logarithms are simply mathematical expressions. For a term x^n in an equation, n is the exponent (or power) and represents the number of times the number x is multiplied by itself. So, "10^3" is the same as $10 \times 10 \times 10 = 1000$, and "$3^5$" is the same as $3 \times 3 \times 3 \times 3 \times 3 = 243$.

A logarithm is simply an inverse exponent. In the term $\log_n x$ the number n is known as the *base* of the logarithm. Taking again the case of 1000 and a base of 10, $\log_{10}(1000) = \log_{10} 10^3 = 3$. Whilst in theory a logarithm can have any base, in practice the bases of most importance are 10 and a number called e. For log to the base e, where $e = 2.718$ (to three decimal places), the logarithm is called the *natural* logarithm and is normally denoted "ln". Logs to the base 10, normally just denoted "log", relate to natural logs (i.e. "ln") by the factor 2.303, i.e. ln(10); so $\ln(x) = 2.303 \times \log(x)$. The number e is very significant, since it defines exponential relationships which are important in things like first-order kinetics (Chapter 5) and chemical reactor theory (Chapter 8).

Table 2.2 Exponent and logarithm rules.

Exponents	Logarithms
$x^0 = 1$	$\log 1 = 0$
$x^1 = x$	$\log_x x = 1$*
$n^x n^y = n^{(x+y)}$	$\log (xy) = \log x + \log y$
$n^x/n^y = n^{(x-y)}$	$\log (x/y) = \log x - \log y$
If $n^{(1/x)} = y$ then $y^x = n$	$\log x^n = n \log x$

*so, $\log_{10} 10 = 1$ and $\log_e e$ (or "ln e") $= 1$

A negative logarithm to the base 10 is denoted p. This term is used in chemistry to define concentration: the *pH* is the negative log of the hydrogen ion concentration in moles per litre, and is a measure of acidity (Equation 4.10). This terminology is also used for any parameter which tends to be defined by orders of magnitude (i.e. "$\times 10^n$", where n in this case is usually negative). Thus if a chemical has an acid dissociation constant K_a (Equation 4.26) of 5×10^{-9}, its pK_a value will be 8.3.

Exponential and logarithmic relationships can be manipulated in similar ways, with five basic rules for each (Table 2.2). Taking logs allows data to be processed more easily because linear equations are produced (i.e. equations taking the general form "$y = mx + c$"). For example, for the equation:

$$y = m x^n \tag{2.1}$$

taking logs produces a linear equation:

$$\log y = \log m + n \log x \tag{2.2}$$

EXAMPLES: MANIPULATING LOGARITHMIC EXPRESSIONS

Look at these following manipulations, and check the answers using a calculator:

(a) $\log 100 = \log (10^2) = 2 \log 10 = 2$
(b) $\log (10^{½} \, 100^{¾}) = ½ \log 10 + (2 \times ¾) \log 10 = (½ + 1½) \log 10 = 2$
(c) $\log \{ (10000^{¾})/100 \} = \log 10^{4 \times ¾} - \log 10^2 = 3 \log 10 - 2 \log 10 = 1$.

Given that $\log 2 \sim 0.3$ and $\log 5 \sim 0.7$, work out the logarithm of the following from (a) manipulating logs, and then (b) checking using a calculator

(d) $2^5/5^2$ Ans. ~ 0.1
(e) 0.5 Ans. ~ -0.3 [NB $0.5 = 5/10$]
(f) 25×8 Ans. ~ 2.3 [NB $25 = 5^2$, $8 = 2^3$].

EXERCISE 2.4

Work out log y in terms of a, b and c for the following expressions, once again assuming log 2 ~ 0.3 and log 5 ~ 0.7.

(a) $c \, 5^{2b}/10^a$
(b) $2^{0.5} \, (0.5^{ab} \, 5^c)$
(c) $2 \, (100^{0.5b} \, 25^{5c}/200^a)$.

Now try the following:

(d) If $a = b^{-m}/(cd^{-m})$, what is (log a) in terms of b, c, d and n?
(e) If $X/X_o = 2^{t/s}$, what is t in terms of s, X, and X_o?

EXAMPLE: DISINFECTION KINETICS

The rate at which a micro-organism is inactivated is described by the equation:

$$N_t = N_0 e^{-kt}$$

where N_0 and N_t define the micro-organism concentrations initially and at time t respectively and k is the rate constant in units of inverse time. If $k = 0.25\ min^{-1}$, how long would it take for the concentration to decline by 99%?

From the above equation:

$$N_t/N_0 = e^{-kt}$$

Taking logs:

$$ln(N_t/N_0) = -kt$$

If N_0 has declined by 99% then $N_t/N_0 = 0.01$. So:

$$t = \frac{-ln(0.01)}{0.25} = 18.4\ \text{minutes.}$$

EXERCISE 2.5
If the pressure in a gas tank decays according to the equation:

$$p = kt^{-n}$$

and k and n are constants, what are the values of these constants if $p = 0.5$ when $t = 3$, and 0.05 when $t = 15$?

2.4 UNITS

Although some design parameters (such as the Reynolds number, see Section 2.4.1) are unitless, most of the parameters important in water and wastewater treatment plant design have specific units. The standard (SI) units used for mass, distance and time are respectively kg, m and s. Most other important parameters are derived from these units alone, as the example below demonstrates. The prefixes of nano, micro and milli respectively represent a billionth (or thousand millionth, 10^{-9}), a millionth (10^{-6}) and a thousandth (10^{-3}) of a quantity. The prefixes kilo, mega and giga represent a thousand, million and billion times the quantity. Key derived basic metric units are the tonne (te), which is 10^3 kg, the gramme (g), which is 10^{-3} kg, and the litre (L), which is 10^{-3} m^3.

EXAMPLE: DETERMINING DERIVED UNITS

(a) *If **force** is **mass** $\times$ **acceleration**, what are the SI units of force (derived SI unit, newton)?*
(b) *If **pressure** is **force per unit area**, what are the SI units of pressure (derived SI unit, pascal)?*
(c) *If **potential energy** is **force** $\times$ **distance**, what are the SI units of energy (derived SI unit, joule)?*
(d) *If **kinetic energy** is also **½ mass** $\times$ **velocity^n**, what is the value of the exponent n?*
(e) *If **power** is **energy per unit time**, what are the SI units of power (derived SI unit, watt)?*

(1) Force = mass (kg) $\times$ acceleration (m/s^2), and so 1 newton (N) = 1 kg $\cdot$ m/s^2
(2) Pressure = force (kg $\cdot$ m/s^2) $\div$ area (m^2), and so 1 pascal (Pa) = 1 kg/(m $\cdot$ s^2)
(3) Energy = force (kg $\cdot$ m/s^2) $\times$ distance (m), and so 1 joule (J) = 1 kg $\cdot$ m^2/s^2
(4) Energy = mass $\times$ velocity^n, and so velocity^n = kg $\cdot$ m^2/s^2 $\div$ kg = m^2/s^2 = v^2
(5) So, $n = 2$ and so kinetic energy = ½ mv^2
(6) Power = energy (kg $\cdot$ m^2/s^2)/time, and so 1 watt (W) = 1 kg $\cdot$ m^2/s^3

EXERCISE 2.6
*(1) What are the units of **energy per unit mass** and **power per unit mass flow**?*
*(2) What parameter has the same units as **energy per unit area per unit velocity**?*

2.4.1 Common parameters

The most common parameters in water and wastewater treatment are length (e.g. diameter and length of pipes or tanks), mass, concentration (in mass or moles per unit volume), pressure and flow (of water or air), and energy and power. It is sometimes required to determine the flow velocity (for example in m/s) from the flow rate (in m^3/h), or the volume (in m^3) from the mass (in kg). Flow velocity is obtained by dividing the volumetric flow rate by the cross-sectional area (in m^2). Mass flow (in kg/m^3) is simply volumetric flow multiplied by the density (in kg/m^3). Finally, derived units are often used, such as megalitres/day (MLD) for flow and litres/($m^2 \cdot$ h) (LMH) for flux through a membrane, since these give convenient numbers.

EXAMPLE: FLOW THROUGH A RECTANGULAR CONDUIT

Water flows through a 1.2 m × 0.5 m rectangular conduit at a velocity of 0.4 m/s. What is the volumetric flow rate in m^3/hr? If the water flows into a pipe what would be the pipe diameter to maintain the same velocity?

The volumetric flow rate (Q) is given by the cross-sectional area (A) and the velocity (U):

$A = 1.2\,m \times 0.5\,m = 0.6\,m^2$

$Q = 0.6\,m^2 \times 0.4\,m/s = 0.24\,m^3/s.$

Converting to m^3/h:

$Q = 3600 \times 0.24 = 864\,m^3/h.$

For a cylindrical pipe the cross-sectional area A is $\pi d^2/4$

So, $\pi d^2/4 = A = Q/U,$

where v must be maintained at 0.4 m/s and Q is 0.24 m^3/s.

Rearranging:

$$d = \sqrt{\frac{4Q}{\pi U}} = 2\sqrt{\frac{0.24}{3.14 \times 0.4}} = 0.874\,m.$$

EXERCISE 2.7
What is the velocity of water flowing at 30 m^3/h through a 50 mm pipe and, at this flow rate, how long would it take to fill a 5 m high cylindrical tank which is 12 m in diameter.

EXERCISE 2.8
Sludge produced from a sewage treatment works has a density of 1020 kg/m^3. If the volume flow is 50 m^3/day what is the mass flow in tonnes/h (te/h)?

Some engineering parameters are unitless. Foremost amongst these is the *Reynolds number (Re)*, which represents the extent to which flow is *turbulent* (Section 3.3) and quantifies the ratio of inertial to laminar flow. It is thus a fundamental parameter defining the *flow regime*. However, there are a significant number of other dimensionless groups applied across a variety of applications or engineering disciplines, all generally representing the ratio of two forces acting on the system (Table 2.3).

Table 2.3 Dimensionless numbers.

Name	Symb.	Formula	Application	Ratio of forces represented
Archimedes	Ar	$gd^3(\rho_p-\rho)\rho/\mu^2$	Particle settling/floating	Buoyancy/inertial
Froude	Fr	v^2/gd	Flocculation	Inertial/gravitational
Peclet	Pe	dv/D	Mixing, mass transfer	Advective/diffusive
Prandtl	Pr	$C_p\mu/k_T$	Mixing, mass transfer	Viscous/thermal diffusive
Reynolds	Re	$\rho vd/\mu$	Flow	Inertial/viscous
Schmidt	Sc	$\mu/\rho D$	Mass transfer	Viscous/mass diffusive
Sherwood	Sh	kd/D	Mass transfer	Convective/diffusive mass transfer
Weber	We	$\rho v^2 d/\sigma$	Bubble drop formation	Inertial/surface tension

g, acceleration due to gravity (m/s^2); d, diameter or characteristic length (m); ρ_p, particle density (kg/m^3); ρ, liquid density (kg/m^3); μ, dynamic viscosity (kg/(ms)); v, velocity (m/s); D, diffusion coefficient (m^2/s); C_p, specific heat capacity (J/(kg · K)); k, mass transfer coefficient (m/s); k_T, thermal conductivity (W/(m.K)); σ, surface tension (kg/s^2)

Table 2.4 Common conversions.

Imperial units	Metric (or pseudo-metric) units	Imperial to metric	Metric to imperial
lbs	kg	0.454	2.20
inches	m	0.0254	39.4
ft	m	0.305	3.28
SCFM	Nm^3/h	1.66*	0.6*
psi	bar	0.0690	14.5
sq ft	sq m	0.0929	10.8
U.S. gal	litres	3.79	0.264
MGD	MLD	3.79	0.264
hp	kW	0.745	1.34
°F	°C	Subtract 32, multiply by 0.55	Multiply by 1.8, add 32

SCFM, Standard cubic feet per minute; *at 21°C; psi, Pounds per square inch; MGD, Megagallons (US) per day; MLD, Megalitres per day; hp, horsepower

---

EXERCISE 2.9
Demonstrate that the Prandtl number is unitless.

2.4.2 Imperial vs. SI units

Whilst the rest of the world employs SI units, or has at least mostly accepted metrification, the US still largely use *Imperial* units, based on length measured in inches and feet, volume in (US) gallons, and mass measured in pounds and tons. This also yields a series of derived units such as psi (pounds of force per square inch for pressure), SCFM (standard cubic feet per minute for gas flow), and hp (horsepower for power) – where one unit of horsepower equates to 550 foot-pounds per second. Clearly, Imperial units are not consistent, and it is normally advisable to convert them to SI units before proceeding with any calculation so as to avoid confusion with the units. There are therefore some key conversion factors needed to convert Imperial quantities to metric ones, a few examples of which are listed in Table 2.4.

Chapter 3

Fluid physics

Water and wastewater treatment relies on the movement of fluids, whether this is the bulk movement of water, addition of dissolved reagents (such as acid, bases and oxidants), or the bubbling of air into a biotank to maintain biotreatment. When water moves through pipes, channels or unit operations, a loss of energy occurs due to friction between the water and the solid surfaces, such as a pipe or granular material in media bed. Since water naturally flows downwards under gravity, wherever possible process treatment schemes are designed to take advantage of this potential energy. However, it is often necessary to pump the fluid. The use of gravity flow and the specification of the pumps demands calculation of system pressure losses to produce a hydraulic profile of the plant, and thereby establish either the height of water and/or the pumping power required.

3.1 PRESSURE

At any depth or height h, in metres, within a fluid a force is exerted over an area A on which the force acts:

$$F = A\rho g h \tag{3.1}$$

where ρ the fluid density in $kg \cdot m^{-3}$ and g the gravitational constant of 9.81 m/s^2. Dividing the force F in *newtons* (N), or $kg \cdot m/s^2$, by the area A gives the *hydrostatic pressure* P_h in N/m^2, kg/(m $\cdot$ s^2) or *pascals* (Pa):

$$P_h = \rho g h. \tag{3.2}$$

Since the pascal is very small, pressures are normally expressed in kPa, MPa or units such as the bar (100,000 Pa), atmosphere (101,325 Pa), m of water (9,806 Pa at 20°C) or mm of mercury (133 Pa at 20°C). Because atmospheric pressure (P_{atm}) acts everywhere, the total (or absolute) pressure is given by $P_h + P_{atm}$, but it is normal when designing process plant to consider only P_h. P_h may thus be positive or negative depending on whether a positive pressure or a vacuum is being applied.

It is often convenient in water engineering to use the depth of water as a measure of pressure, or *head* of water since, according to Equation 3.2:

$$h = P_h/(\rho g). \tag{3.3}$$

Since water has a density of $\sim$1000 kg/m^3 a pressure of 1 bar is about 10 m head of water, as compared to only 750 mm of mercury (chemical symbol Hg) which has a considerably higher density. Pressure in "mm Hg" is still sometimes used, since barometers used to be based on liquid mercury.

© IWA Publishing 2019. Watermaths: Process Fundamentals for the Design and Operation of Water and Wastewater Treatment Technologies
Author: Simon Judd
doi: 10.2166/9781789060393_0017

3.2 VISCOSITY

Viscosity is a measure of how readily fluids flow, and arises largely from their molecular size and structure. Fluids like water, alcohol or petrol which flow readily have low viscosity, whilst liquids like treacle, glycerine and glues have high viscosities. In fluid mechanics terms, the viscosity relates the *rate of deformation* or *velocity gradient* of the fluid, dv/dh, to an applied *shear stress*, τ, in $kg/(m \cdot s^{-2})$, by *Newton's Law*:

$$\tau = \mu(dv/dh) \tag{3.4}$$

where μ = absolute viscosity in $N \cdot s/m^2$ or $kg/(m \cdot s)$ or *poise* (which is $0.1\,kg/(m \cdot s)$). The velocity gradient dv/dh, sometimes called the *shear rate*, is the change in velocity, v, with liquid depth, h. Viscosity decreases with increasing temperature (Table 3.5); at 20°C the value for water is $\sim 10^{-3}\,kg/(m \cdot s)$ and for air at the same temperature it is $1.8 \times 10^{-5}\,kg/(m \cdot s)$. Fluids like water and air, which obey Newton's Law, are referred to as *Newtonian* but many fluids are *non-Newtonian*, in that the viscosity either increases or decreases with the shear rate or with time. Sewage sludge, for example, behaves like a solid at low shear stresses but flows as a viscous fluid at high shear stresses.

It is sometimes convenient in fluid mechanics calculations to use the ratio of absolute viscosity to density to produce the *kinematic viscosity* v, which takes units of m^2/s or *Stokes*, where 1 Stokes $= 10^{-4}\,m^2/s$:

$$v = \mu/\rho. \tag{3.5}$$

3.3 BOUNDARY LAYER AND THE REYNOLDS NUMBER

If a fluid flows at $Q\,m^3/s$ in a pipe of cross section area $A\,m^2$, the bulk velocity, v, is:

$$v = Q/A. \tag{3.6}$$

However, since the fluid is viscous, the fluid close to the pipe wall moves more slowly than this, and at low to moderate flows a *velocity profile* exists across the pipe (Figure 3.1). The slow moving layer of fluid close to the wall is called the *boundary layer* and its thickness is defined by the point at which the velocity becomes equal to 99% of the free stream flow.

If the free stream velocity is increased this orderly *laminar flow* pattern begins to break up and the flow becomes *turbulent*. The change in flow regime is defined by the *Reynolds Number, Re*, which for a pipe of internal diameter d is defined as:

$$Re = \rho vd/\mu = vd/v \tag{3.7}$$

where flow at $Re \lesssim 2000$ is regarded as laminar and is mostly turbulent at $\gtrsim 4000$. Between these two limits the flow is termed *transitional*. The Reynolds number is a key parameter in fluid mechanics calculations.

EXERCISE 3.1

What is the Reynolds number of water flowing through a 50 mm diameter pipe at 25 m^3/h, assuming the density and viscosity are 1000 kg/m^3 and 0.001 $kg/(m \cdot s)$ respectively? By how much would this change if the water temperature was 10°C?

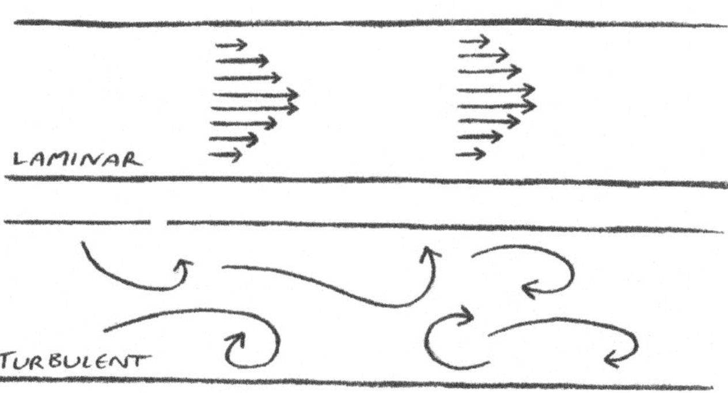

Figure 3.1 Laminar vs. turbulent flow.

3.4 BERNOULLI'S EQUATION

For an elevated tank containing water and fitted with a pipe at its base discharging at ground level, the water has potential energy on the basis of its height which is defined as the *static head H* in metres of water (Figure 3.2). If the tank is sealed and pressurised the additional energy is referred to as the *pressure head $P/(\rho g)$*, where P is pressure in pascals. When the water is allowed to flow from the tank along the pipe the potential energy is converted into kinetic energy, or the *velocity head*, $v^2/2\,g$. Given that energy is always conserved, the sum of these three quantities is constant. This is the basis of *Bernoulli's equation*:

$$H + P/(\rho\,g) + v^2/(2\,g) = \text{constant} \tag{3.8}$$

Bernoulli's equation can be used to determine the pressure for a given system flow, or vice versa. The example of the elevated tank is provided below. In practice, however, the flows calculated from this equation are reduced by the impact of frictional forces acting at the water–pipe interface. These are accounted for by calculating the friction losses.

EXAMPLE: FLOW FROM A TANK

Water at 20°C in a sealed tank 6 m above ground level pressurised with compressed air to 1.5 bar (15 m) is then discharged at a height of 1.5 m above ground through a 75 mm diameter pipe (Figure 3.2). How fast does the water flow through the pipe?

Initially the water is static and so the velocity head is zero, the static head is 6 m and its pressure head 15 m. At the end of the pipe the pressure is atmospheric and the static head is 1.5 m. An energy balance based on Equation 3.8, shows indicates:

$$6 + 15 + 0 = 1.5 + 0 + v^2/(2 \times 9.81)$$

So, $v^2 = 2 \times 9.81 \times (6 + 15 - 1.5)$

$$v = 19.6\,\text{m/s}$$

and so the flow rate (Q) in the 75 mm pipe (cross-sectional area of $(\pi/4) \times 0.075^2 = 0.00442\ \text{m}^2$) would be:

$$Q = 19.6 \times 0.00442 = 0.0866\,\text{m}^3/\text{s, or } 3.12\,\text{m}^3/\text{h}.$$

EXERCISE 3.2

What would be the pipe velocity if the tank in Figure 3.2 were to be raised a further 6 m and opened to the atmosphere? How much pressure would be required to fill the (unpressurised) tank if the water had to be taken a river 2.5 m below ground level?

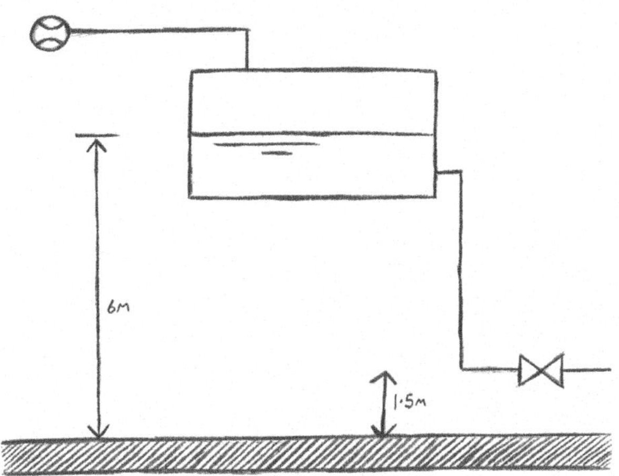

Figure 3.2 Flow from a pressurised tank.

Table 3.1 Pipe roughness values in mm.

Pipe characteristics	κ (mm)
New plastic and non-ferrous	0.03
Spun bitumen or cement lined ductile iron	0.05
Steel (uncoated)	0.05
Good ductile or cast iron	0.05
Galvanised steel	0.15
Precast concrete	0.15
Tuberculated water mains up to 20 years old	1.5–15
Tuberculated water mains up to 50 years old	0.3–30

3.5 FRICTION LOSSES IN PIPES AND FITTINGS

3.5.1 Pipes

When water flows through a pipe the friction between the wall of the pipe and the water slows the water down and uses up some of its energy. As a result a loss of head ΔH (normally denoted H_f) along the pipe results, which relates to the pipe diameter d and length l and the free stream velocity v by the *Darcy–Weisbach* equation:

$$H_f = \lambda \frac{l}{d} \frac{v^2}{2\,g} \tag{3.9}$$

where λ is the *Darcy–Weisbach friction factor*, and depends on the surface roughness, κ, and the Reynolds number. It also usual practice to define the head loss per unit length of pipe ($\Delta H/l$) as the *hydraulic gradient*, s, so that the above equation becomes:

$$s = \frac{\lambda}{d} \frac{v^2}{2\,g} \tag{3.10}$$

λ can be obtained from *Moody diagrams*, derived from the *Colebrook–White* equation. The Colebrook–White equation is not readily employed directly, but several simplifications exist, including the *Swamee–Jain* approximation:

$$\lambda = \frac{0.25}{[\log(((\kappa)/(3.7d)) + ((5.74)/(Re^{0.9})))]^2} \tag{3.11}$$

where pipe roughness (κ) takes units of length and typical values are given in Table 3.1. Under highly turbulent conditions, i.e. when $5.74/Re^{0.9} \ll \kappa/3.7D$, the second term in the denominator of Equation 3.11 disappears and the expression simplifies to:

$$\lambda = \frac{0.25}{[\log((\kappa)/(3.7d))]^2} \tag{3.12}$$

EXAMPLE: HEAD LOSS ALONG A PIPE

At what rate does the water flowing from the pressurised tank shown in Figure 3.2 flow along a 20 m long 75 mm diameter galvanised steel pipe?

At the original flow velocity of 19.6 m/s, and a viscosity of 10^{-3} kg/(m · s) at the temperature of 20°C (Table 3.5) the Reynolds number (Equation 3.7) can be calculated as:

$$Re = 1000 \times 19.6 \times 0.075/10^{-3} = 1.47 \times 10^6$$

According to Table 3.1 the κ value for galvanised steel can be assumed to be 0.15 mm. Calculating the two terms in the denominator of Equation 3.11 reveals that the first term is given by:

$$\kappa/(3.7\,d) = 0.15/(3.7 \times 75) = 5.41 \times 10^{-4}$$

whereas the second term, containing the Reynolds number, is:

$$5.74/Re^{0.9} = 1.62 \times 10^{-5}$$

So, from Equation 3.11:

$$\lambda = \frac{0.25}{[\log\ (5.41 \times 10^{-4} + 1.62 \times 10^{-5})]^2} = \frac{0.25}{[\log(5.57 \times 10^{-4})]^2} = 0.0236.$$

From Equation 3.10, the hydraulic gradient, s, associated with flow through the pipe at this friction factor value is given by the velocity head term:

$$s = v^2\, 0.0236/(0.075 \times 2 \times 9.81) = 0.0160\, v^2 \text{ m head per m pipe length}$$

So, for a 20 m pipe length total head loss through the pipe is $20 \times 0.0160 = 0.320\, v^2$ m.

Substituting this figure into the Bernoulli Equation 3.8 as before yields:

$$6 + 15 + 0 = 1.5 + 0 + v^2/(2 \times 9.81) + 0.320\, v^2 = 1.5 + v^2(0.051 + 0.320)$$

So, $\quad v^2 = (6 + 15 - 1.5)/(0.320 + 0.051) = 52.6$

$$v = 7.25\,\text{m/s}.$$

The cross-sectional area of the 75 mm pipe is $(\pi/4) \times 0.075^2 = 0.00442\ \text{m}^2$. So, the flow rate is:

$$Q = 7.25 \times 0.00442 \times 3600\ \text{m}^3/\text{h} = 115\ \text{m}^3/\text{h}.$$

Thus, frictional losses in the pipe reduce the flow in this instance from 311 to 116 m^3/h – a reduction of 63%. However, this calculation assumes that the revised flow does not impact on λ, the friction factor, through the Reynolds number term in Equation 3.11. Recalculating:

$$Re = 1000 \times 7.25 \times 0.075/10^{-3} = 5.44 \times 10^5$$

Thus $\quad 5.74/Re^{0.9} = 3.95 \times 10^{-5}$.

Reinserting this value into Equation 3.11 produces a revised λ value:

$$\lambda = \frac{0.25}{[\log\ (5.41 \times 10^{-4} + 3.95 \times 10^{-5})]^2} = \frac{0.25}{[\log\ (5.81 \times 10^{-4})]^2} = 0.0239.$$

This is only 1% different to the λ value of 0.0236 calculated from the original velocity, an insignificant change which can be ignored and the flow taken as 115 m^3/h.

EXERCISE 3.3
Estimate the head loss along 2 km of 300 mm diameter ductile iron pipe for water flowing at 320 m^3/h.

3.5.2 Pipe fittings

Losses occur not just in the lengths of pipe but also at bends, constrictions, valves, T-pieces and orifice plates. These head losses all relate to the velocity head term of the Bernoulli equation, and for each individual fitting:

$$H = k_{\text{fitting}} v^2/(2\,g) \tag{3.13}$$

where the coefficient k_{fitting} takes an empirical (i.e. practically determined) value specific to the fitting (Table 3.2). Given that losses are additive, the total head loss is simply that given by the sum of all the k_{fitting} values making up the velocity head.

Table 3.2 Fitting coefficient values for head loss.

Fitting	k_{fitting}
Sharp edged entrance	0.50
Bellmouth entrance	0.05
Close radius 90° bend	0.75
Close radius 45° bend	0.30
Long radius 90° bend	0.40
Long radius 45° bend	0.20
Tee with flow in line	0.35
Tee with flow in branch	0.80
Reducer	0.15
Sudden contraction 3:1*	0.45
Sudden contraction 5:1*	0.50
Sudden enlargement 1:3*	0.80
Sudden enlargement 1:5*	1.00
Bellmouth exit	0.20
Gate valve fully open	0.12
Butterfly valve fully open	0.45
Reflux valve	1.00

*For a change from a pipe diameter 1 to diameter 2, k_{fitting} is given by $(v_1 - v_2)^2/2\,g$.

EXAMPLE: HEAD LOSS THROUGH FITTINGS

What would be the adjusted velocity if the 20 m pipe from Figure 3.2 had six close-radius 90° bends as well as a sharp-edged entrance at the tank outlet and a sudden enlargement at the end of the pipe?

From Table 3.2, the sum of the k_{fitting} values for the six close radius bends, the sharp-edged entrance and the sudden enlargement (assuming a 1:3 ratio) is:

$$\Sigma k_{\text{fitting}} = 6 \times 0.75 + 0.5 + 0.8 = 5.8$$

So, from the Bernoulli equation once again the total head loss provided by the fittings is:

$$H = 5.8\,v^2/(2g)$$

Applying the same Bernoulli balance as in Sections 3.4 and 3.5.1, and recognising that both the inlet velocity and the outlet applied pressure are zero:

$$6 + 15 + 0 = 1.5 + 0 + \frac{v^2}{2\,g} + H_{f,\text{pipe}} + H_{f,\text{fittings}};$$

$$19.5 = \frac{v^2}{2\,g} + H_{f,\text{pipe}} + H_{f,\text{fittings}} = H_{f,\text{pipe}} + \frac{v^2}{2\,g}(5.8 + 1)$$

where, from Equation 3.9:

$$H_{f,\text{pipe}} = \lambda \frac{l}{d} \frac{v^2}{2\,g} = \lambda \frac{20}{0.075} \frac{v^2}{2\,g} = 267\lambda \frac{v^2}{2\,g}$$

Thus $\quad 19.5 = \frac{v^2}{2\,g}(267\lambda + 6.8);$

So $\quad v^2 = 2g \times 19.5/(267\lambda + 6.8) = 383/(267\lambda + 6.8)$

So $\quad v = \left(\dfrac{383}{267\lambda + 6.8}\right)^{0.5}$

(A)

However, according to Equation 3.11, λ is a function of the Reynolds number and so velocity, and hence is also unknown:

$$\lambda = \frac{0.25}{\left[\log\left(\dfrac{\kappa}{3.7d} + \dfrac{5.74}{Re^{0.9}}\right)\right]^2}, \quad \text{where } Re = \frac{\rho U d}{\mu} \tag{B}$$

Equations A and B must be solved by iteration, i.e. repeating the calculation until the value of v converges to a single value. To start this, a value of λ must be selected and used to calculate v from Equation A, and the resultant value used to calculate Re and thus λ using Equation B. This is best achieved through spreadsheeting, but to demonstrate manually, using the value of λ for 0.0236 taken from the first example of flow through a frictionless pipe:

$$v = \left(\frac{383}{267 \times 0.0236 + 6.8}\right)^{0.5} = 5.41\,\text{m/s},$$

So $Re = 1000 \times 5.41 \times 0.075/0.001 = 4.06 \times 10^5$.

Entering this into Equation B:

$$\lambda = \frac{0.25}{\left[\log\left(\dfrac{0.15}{3.7 \times 75} + \dfrac{5.74}{406,000^{0.9}}\right)\right]^2} = \frac{0.25}{[\log\,(5.41 \times 10^{-4} + 5.14 \times 10^{-5})]^2} = 0.0240.$$

Re-entering this value into Equation A produces:

$$v = \left(\frac{383}{267 \times 0.024 + 6.8}\right)^{0.5} = 5.38\,\text{m/s}$$

This value differs by less than 1% from the previously determined value and so can be considered representative.

EXERCISE 3.4
What would be the new pipe velocity for the set-up shown in Figure 3.2 if the pipe was 100 mm diameter uncoated steel and all six bends were long radius?

EXERCISE 3.5
A submersible pump delivers water from a borehole at water level 25 m below ground level to a tank at 2.5 m above ground level. The pump is fitted with a gate valve for isolation and the 250 m long, 100 mm diameter ductile iron pipe includes six long radius 90° bends. The inlet to the tank can be regarded as a sudden enlargement (1:5). What is the required pump delivery head at a flow of 43 m^3/h?

Generally, liquid flow velocities in pipework are in the range 0.5–2.5 m/s to limit the head loss to, typically, 0.01–1.0 m per 100 m. Large pipes are more expensive than small ones but give a lower head loss so there is a trade-off between capital cost of the pipe and operating cost (pumping power) of transmitting the water. Greater velocities may be needed where levels of settleable suspended solids are high, since these may otherwise be deposited in the pipework.

In addition to the fittings shown in Table 3.2, there are also orifice plates (a plate inside the pipe with a hole drilled in the centre) which may be introduced to control flow by providing a head loss. An orifice of cross-sectional area a in a pipe of cross-section A generates a head loss of H metres at an in-hole velocity v_{hole} according to the expression:

$$v_{\text{hole}} = J\sqrt{\frac{2gH}{(1 - ((a^2)/(A^2)))}} \tag{3.14}$$

where J is a coefficient of value 0.61–0.62.

EXAMPLE: HEAD LOSS THROUGH AN ORIFICE

What would be the head loss through a 15 mm diameter orifice in a 50 mm diameter pipe flowing at 8 m³/h?

Flow $Q = 8\,\text{m}^3/h = 8/3600 = 2.22 \times 10^{-3}\,\text{m}^3/s$
Orifice diameter $= 15/1000 = 0.015\,\text{m}$
So, $a = \pi d^2/4 = 3.14 \times 0.015^2 \times 0.25 = 0.177 \times 10^{-3}\,\text{m}^2$

So, in − hole velocity $v = Q/a = 2.22/0.177 = 12.5\,\text{m/s}$

From Equation 3.14, $v_{\text{hole}} = C\sqrt{[2gH/(1 - a^2/A^2)]} = C\sqrt{[2gH/(1 - d_{\text{hole}}^4/d^2)]}$.

where $(1 - d_{\text{hole}}^4/d^4) = 1 - (d_{\text{hole}}/d)^4 = 1 - (15/50)^4 = 1 - 0.0081 = 0.992$

Assuming $C = 0.61$, and substituting into Equation 3.14:

$12.5 = 0.61\sqrt{(2 \times 9.81 \times H/0.992)}$

So, $H = (12.5/0.61)^2 \times 0.992/(2 \times 9.81) = 21.2\,\text{m}$

3.6 FLOW IN OPEN CHANNELS

Friction losses in open channels are less easily calculated than those in pipes, but are usually low (typically <0.1 m head per 100 m channel length) and can be found in published tables and nomograms. An example of a simplified relationship used is the *Manning–Strickler* formula which relates the flow velocity to wall roughness K (Table 3.3), mean hydraulic depth R (the cross-sectional area of the liquid in the channel in m² divided by the wetted perimeter in m) and channel slope s (in m height per m length):

$$v = KR^{0.66}s^{0.5}. \tag{3.15}$$

Head losses of H can be caused by flow through a submerged horizontal slot or *penstock* (Figure 3.3), formed by a *baffle*, or wall that fills part of a channel. At flow Q m³/h and penstock height h and width w in metres the relationship is roughly given by:

$$Q \sim jhw\sqrt{(2gH)} \tag{3.16}$$

where J is the coefficient as before (i.e. taking a value of 0.61–0.62).

Table 3.3 Manning–Strickler wall roughness values.

Channel characteristics	K
Smooth cement rendered	60
Vitrified clay	50
Concrete	40
Metal flume	30
Earth bank	20
Grassy bank	10

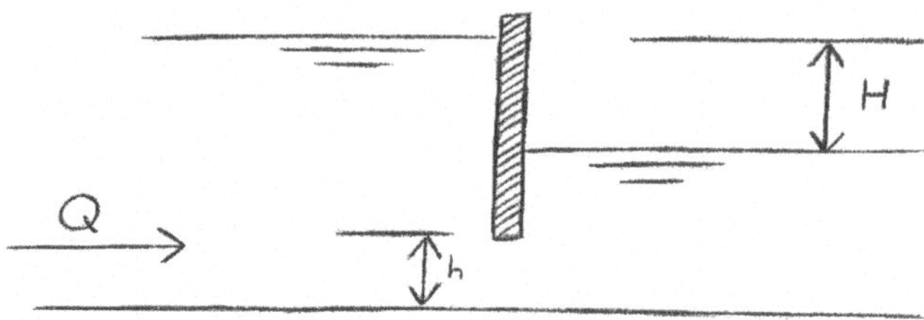

Figure 3.3 Flow under a baffle, or penstock.

EXAMPLE: HEAD LOSS THROUGH A PENSTOCK

Calculate the head loss through a 500 mm square penstock at a flow of 1120 m³/h.

Flow rate $Q = 1120/3600 = 0.311\,\mathrm{m^3/s}$
assume $C = 0.61$
$\quad d = w = 500/1000 = 0.5\,\mathrm{m}.$
According to Equation 3.16:

$\quad Q \sim Chw\sqrt{(2gH)}$

so, $0.311 = 0.61 \times 0.5\ \times 0.5 \times \sqrt{(2 \times 9.81 \times H)}$
thus $\sqrt{(19.6H)} = 2.04$

$\quad H = 2.04^2/19.6 = 0.212\,\mathrm{m} = 212\,\mathrm{mm}.$

Head losses caused by vertical slots, either formed between vertical baffles or between one vertical baffle and the vertical wall of the channel, are less easily calculated than those caused by horizontal slots. Firstly, if the total width of the flow channel is B, and the slot width (baffle-to-wall or baffle-to-baffle) is b, then the velocity v in the slot can be calculated from the flow Q and the liquid cross-sectional area A:

$$v = (Q/A) \times (B/b). \tag{3.17}$$

A factor, M, is then calculated from:

$$M = 1 - (b/B). \tag{3.18}$$

From the calculated value of M two further factors, J_1 and J_2, are obtained empirically (Table 3.4). The head loss is then approximately given by:

$$H \sim J_1 \frac{v^2}{2\,gJ_2^2} \tag{3.19}$$

Serpentine flows created by a number of baffles (Figure 3.4) are subject to the same head loss whether the flow is over and under the baffles or from side to side. If v_1 is the approach velocity (i.e. the velocity of the liquid travelling horizontally between the baffle and the wall) and v_2 the velocity between the baffles (i.e. the velocity orthogonal to the overall flow direction), and N the total number of baffles, then:

$$H \sim (1/2\,g) \times \{Nv_1^2 + (N - 1)v_2^2\} + \text{channel friction} \tag{3.20}$$

where the channel friction head loss is given by Equations 3.9−3.11.

Table 3.4 J_1 and J_2 values.

M	J_1	J_2
0.9	0.59	0.97
0.8	0.59	0.80
0.7	0.60	0.73
0.6	0.60	0.65
0.5	0.63	0.58
0.4	0.66	0.50

Figure 3.4 Serpentine flow through baffles.

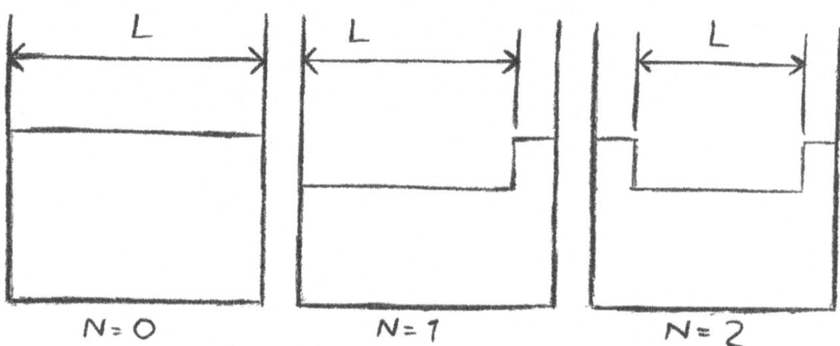

Figure 3.5 Sharp-crested rectangular weirs, as seen from the front, illustrating the value taken for N.

Sharp-crested rectangular weirs, or weirs forming a tank end wall over which water flows, are defined by the number of ends (N in Figure 3.5). The head loss H at flow Q and effective weir length L is given by the *Francis* formula:

$$Q = K(L - 0.1\ NH)\ H^{1.5} \tag{3.21}$$

where K is a function of the drag coefficient and is normally taken as 1.84.

EXAMPLE: HEAD LOSS THROUGH A SLOT

What is the head loss through a 300 mm × 200 mm slot at a flow of 50 m^3/h?

Flow rate $Q = 50/3600 = 0.0139$ m^3/s.

For slot dimensions of $d = 0.2$, and $w = 0.3$ m, according to Equation 3.16:

$Q \sim Chw\sqrt{(2gH)}$.

Assuming $C = 0.61$:

$0.0139 = 0.61 \times 0.2\ \times 0.3 \times \sqrt{(2 \times 9.81 \times H)}$.

So, $\sqrt{(2 \times 9.81 \times H)} = 0.379$

Thus $H = 0.379^2/(2 \times 9.81) = 7.35 \times 10^{-3}$ m $= 7.4$ mm.

EXAMPLE: HEAD LOSS OVER A WEIR

Calculate the head loss when 900 m^3/h of water flows over a 1.8 m long sharp-crested rectangular weir 0.2 m from the vertical channel wall.

$Q = 600\ m^3/h = 900/3600 = 0.25\ m^3/s$.

According to Equation 3.21, the loss H over the weir is:

$Q = K(L - 0.1NH) \times H^{1.5}$

where $L = 1.8$ m, $Q = 0.25$ m^3/s, and K is assumed to be 1.84.
In this case $N = 1$ (Figure 3.5) and so:

$0.25 = 1.84\ [1.8 - (0.1 \times 1 \times H)] \times H^{1.5}$

This equation can only be solved iteratively (Section 6.2.2), but if it can be assumed that $0.1\ H \ll 1.8$ (i.e. $H \ll 1.8$ m), then the equation simplifies to:

$0.25 = 1.84 \times 1.8 \times H^{1.5}$

And so $H = [0.25/(1.84 \times 1.8)]^{1/1.5} = 0.238$ m, which is only 1.3% of 18 m and thus the above assumption is valid.

EXERCISE 3.6

What is the minimum slope required on a 500 mm wide smooth cement rendered channel to maintain a velocity of 1.5 m/s at a flow of 300 m^3/h?

EXERCISE 3.7

A rectangular concrete contact tank treating 72 m^3/h of water is 1.0 m deep and is divided by 15 vertical baffles to form a serpentine path 48 m long by 0.25 m wide with a full width (0.25 m) weir at each end. If the inlet weir level is 1.0 m above the tank floor what must be the height of the outlet weir to ensure flow through the tank?

3.7 FLOW THROUGH POROUS MEDIA

A porous medium is created by a bed of granular material such as sand, activated carbon or sludge solids, spherical beads such as ion exchange media, or fibrous mats. Flow through these media is nearly always vertical – either upflow or downflow – and the medium exerts a head loss due to the resistance to flow it offers. This resistance relates mainly to the size of the media material (such as the sand grain diameter) and the gaps between them, and yields head losses generally between 1 and 10 m for clean beds which increase by 50–100% during a typical filter run as the channels become partially blocked.

The simplest expression for describing head loss H through a homogenous porous medium of length l when water at an approach velocity v_a (i.e. volumetric flow rate/gross cross-sectional area) flows through it is *Darcy's Law*, which is valid for laminar flow:

$$v_a = \sigma H/l = \sigma s \tag{3.22}$$

where σ is the permeability of the medium in m/s and depends on fluid characteristics (viscosity, density) and the characteristics of the porous material, which may change with bed depth but are constant for a given medium and temperature. The permeability σ of media used for filtration varies between $\sim10^{-2}$ and $\sim10^{-5}$ m/s depending on the media bed physical properties, and primarily the porosity or voidage ε and particle size d. The porosity is the fraction of the bed which is occupied by the water. The relationship between the hydraulic gradient and these properties is given by the *Kozeny–Carman* equation:

$$s = 5\frac{\mu\, v_a}{\rho\, g}\frac{(1-\varepsilon)^2}{\varepsilon^3}\left(\frac{6}{d}\right)^2 \tag{3.23}$$

where μ and ρ are the water viscosity and density as before, and are temperature dependent (Table 3.5), and g is acceleration due to gravity. This equation is really only valid for spherical grains and homogeneous media, for which the porosity is around 0.4. For a given granular medium ε may be anywhere between 0.35 and 0.6, but will naturally be lower for a bed of mixed particle sizes due to the filling of the voids from the larger particles by the smaller ones.

In practice filter beds usually stratify when backwashed, with the larger, denser particles sinking to the bottom of the bed and the smaller and less dense ones rising to the top. The hydraulic gradient consequently changes with bed depth. Moreover, all

Table 3.5 Density and viscosity of water.

Temperature, °C	Density, ρ, kg/m^3	Viscosity, μ, kg/(m·s)
10	999.73	1.3097×10^{-3}
15	999.13	1.1447×10^{-3}
20	998.23	1.0087×10^{-3}
25	997.07	0.8949×10^{-3}
30	995.68	0.8007×10^{-3}
35	994.06	0.7225×10^{-3}
40	992.25	0.6560×10^{-3}
45	990.25	0.5988×10^{-3}
50	988.07	0.5494×10^{-3}

Table 3.6 Head loss coefficients for various media.

Medium	Size, d (mm)	k_h
Ion exchange resin	0.4–0.5	0.12
Activated carbon	0.6–0.7	0.03
16/30 mesh sand	0.5–1.0	0.06
14/25 mesh sand	0.6–1.2	0.05
8/16 mesh sand	1.0–2.0	0.04
Grade 2 anthracite	1.2–2.5	0.01

granular solids undergo compaction to varying degrees, with compaction increasing with increasing run time and increasing head loss.

Given that the Kozeny–Carman equation is based on ideal grains, it is more accurate to use empirical measurements of the permeability and these are normally expressed as the head loss coefficient k_h (Table 3.6), where over a reasonable range of approach velocity values v_a of 1–20 m/h the hydraulic gradient is given by:

$$s = 1.02^{(15-T)} v_a k_h.$$

(3.24)

where T is the temperature in °C and v_a is the approach velocity in m/h.

EXAMPLE: KOZENY–CARMAN EQUATION

A waterworks sand filter uses a 700 mm deep bed of sand of effective grain diameter 0.55 mm and 0.40 bed porosity. If the approach velocity is 7.5 m/h what is the head loss through the clean sand bed at 20°C, according to the Kozeny–Carman equation?

According to Equation 3.23, the hydraulic gradient is given by:

$$s = \frac{H}{L} = 5 \frac{\mu \, v_a}{\rho \, g} \frac{(1-\varepsilon)^2}{\varepsilon^3} \left(\frac{6}{d}\right)^2$$

where

L = filter depth = 700 mm = 0.7 m
v_a = approach velocity = 7.5 m/h = 7.5/3600 = 2.08×10^{-3} m/s
μ = viscosity at 20°C = 1.009×10^{-3} kg/(m · s) according to Table 3.5
ρ = density at 20°C = 998.23 kg/m^3 according to Table 3.5
g = acceleration due to gravity = 9.81 m/s^2
ε = porosity = 0.4
d = grain diameter = 0.55 mm = 5.5×10^{-4} m.

So, head loss is:

$H = 0.7 \times 5 \times (1.009 \times 10^{-3}/998.23) \times (2.08 \times 10^{-3}/9.81) \, [(1 - 0.4)^2/0.4^3] \times (6/5.5 \times 10^{-4})^2$
$H = 0.502$ m.

EXERCISE 3.8
By how much would the head loss from the above example change if the water temperature dropped by 5°C, the bed compacted by 10%, the grain size decreased by 15% and the approach velocity increased by 20%?

EXERCISE 3.9
What is the total head loss from pumping water at 12°C through a 1.8 m diameter, 1.5 m deep sand bed of 14/25 mesh sand at flow rate of 30 m^3/h.

Chapter 4
Chemical stoichiometry and equilibria

4.1 INTRODUCTION

Chemistry, as it relates to water and wastewater treatment, concerns the reaction or interaction between species in water on an ionic or molecular basis. For example, water itself can be formed from an explosive chemical reaction between hydrogen and oxygen. Two hydrogen molecules react with one oxygen molecule to form two molecules of water, so the molecular ratio for the reaction (or *stoichiometry*, Section 4.5) is 2:1. However, since an oxygen molecule weighs a lot more than a hydrogen molecule the weight ratio is 1:8. Chemistry also encompasses the balance between the reactants and products when the chemical reaction has reached steady state (an area of physical chemistry called *equilibrium thermodynamics*, Section 4.6), and rates of reactions (or *kinetics*, Chapter 5).

Chemical reactions, and thus treatment methods involving chemicals, are often sensitive to chemical conditions of the water. Of these, often of key importance are water quality parameters such as temperature, salinity and, most commonly, acidity expressed as *pH* (Section 4.4). The pH can impact on reactivity (for example of chemical oxidants like chlorine), solubility (of both salts and gases) and chemical equilibria generally such as that which exists for bicarbonate (Section 4.6.4). Chemistry is thus fundamentally based on a consideration of all components of a system on an atomic level. This demands firstly that the amounts of pollutants and reagents in water are considered on this basis.

4.2 ATOMS, MOLECULES AND IONS, AND CHEMICAL WEIGHTS

Matter is made of *atoms*, which are indestructible under normal conditions, and a material made up of only one type of atom is called an *element*. The names and corresponding chemical symbols– in effect, shorthand notation – of the elements commonly arising in water and wastewater treatment are listed in Table 4.1; many other elements exist, some being of importance as pollutants. The table includes data for the *atomic weight*, or the relative weight of atoms of the elements.

The relative atomic weight of any element is normally a whole number. This is because the main sub-atomic particles of *protons*, *neutrons* and *electrons* respectively have relative weights of 1, 1 and 0, and corresponding relative electrical charges of $+1$, 0 and -1. An atom can have any number of protons and neutrons, but for electroneutrality to be preserved the number of electrons must equal the number of protons. It stands to reason, therefore, that the simplest atom will have one proton and one electron, and this is the case for hydrogen. Hydrogen (H) thus has a relative weight of one, and all other elements have atomic weights which are multiples of the weight of a hydrogen atom. This multiple is referred to as the element's *relative atomic weight*. Where there are exceptions, such as chlorine (Cl), it is because the element is made up of a natural blend of *isotopes* – atoms having two different weights due to differing numbers of neutrons but otherwise displaying the same chemical characteristics. Chlorine is actually a 75:25 blend of isotopes with relative atomic weights of 35 and 37, and its relative atomic weight is thus 35.5.

© IWA Publishing 2019. Watermaths: Process Fundamentals for the Design and Operation of Water and Wastewater Treatment Technologies
Author: Simon Judd
doi: 10.2166/9781789060393_0029

Table 4.1 Elements of interest in water and wastewater treatment.

Name	Symbol	Atomic weight
Aluminium	Al	27
Calcium	Ca	40
Carbon	C	12
Chlorine	Cl	35.5
Fluorine	F	19
Hydrogen	H	1
Iron	Fe	56
Magnesium	Mg	24
Manganese	Mn	55
Nitrogen	N	14
Oxygen	O	16
Phosphorus	P	31
Sodium	Na	23
Sulphur	S	32

To become more stable (or less reactive), atoms may either combine with others to form *molecules* or can gain or lose charge (as electrons) to become *ions*. Molecules may also react to form ions or other products, depending on the relative stability of the reactant(s) and product(s). These reactions, and particularly those for some carbon-based organic compounds (such as natural organic matter in the environment), can be very complicated involving a number of individual steps between the combining of the reactants and the formation of the products. However, for the purposes of water and wastewater treatment process design and operation, the key aspects of chemical reactions are:

(1)   the concentration of the chemical species in the water on a *molar* basis
(2)   the involvement of a change in *oxidation state* as part of the chemical reaction.

The symbol for an ion is the appropriate chemical symbol plus the number and sign of charges it has acquired superscripted. In cases where there is more than one atom in the ion or molecule this number is subscripted (Table 4.2). Positively charged ions

Table 4.2 Common ions in water and wastewater treatment.

Ion name	Symbol	Ionic weight	Equivalent weight
Aluminium	Al^{3+}	27	9
Ammonium	NH_4^{+}	18	18
Bicarbonate	HCO_3^{-}	61	61
Calcium	Ca^{2+}	40	20
Chloride	Cl^{-}	35.5	35.5
Ferric	Fe^{3+}	56	18.7
Ferrous	Fe^{2+}	56	28
Fluoride	F^{-}	19	19
Hydrogen	H^{+}	1	1
Hydroxyl	OH^{-}	17	17
Magnesium	Mg^{2+}	24	12
Nitrate	NO_3^{-}	62	62
Nitrite	NO_2^{-}	46	46
Phosphate	PO_4^{3-}	95	31.7
Sodium	Na^{+}	23	23
Sulphate	SO_4^{2-}	96	48

are called *cations* and negatively charged ones are *anions*. For example the sodium ion is a Na atom which has lost an electron to become Na^+, the sodium cation, and the sulphate ion is the group SO_4 which has gained two electrons to become SO_4^{2-}, the "sulphate" anion. In both cases, the constituent atoms become stable through adopting a charge: the reaction of elemental sodium with water is actually very violent since sodium ions are much more stable than the metal. The chemical reaction involves the transfer of electrons from one reacting species to the other, producing a *redox* reaction (Section 4.5).

The number of charges on an ion is sometimes referred to as the *valency*: hence, the sodium ion has a valency of $+1$ whilst the sulphate ion has a valency of -2. However, valency can also refer to bonding within an organic molecule (so-called *covalent bonding*). Covalent bonding is important in reactions involving organic compounds, such as the chlorination of humic substances (the natural organic matter from soil, vegetation, etc).

Table 4.2 lists the most common ions in water. Some ions have name different from the parent element. For example Cl^- is always called *chloride*, as opposed to the *chlorine* atom Cl or chlorine gas molecule Cl_2. Some elements such as iron can form ions with different charges, e.g. the iron cations *ferric* (Fe^{3+}) and *ferrous* (Fe^{2+}), which have quite different chemical properties from one another and so demonstrate different behaviour in water.

Table 4.2 includes both the *ionic weight* and the *equivalent weight* of the ion. The ionic weight is the combined weight of the atoms making up the ion: the charge has no impact on the weight since it results only from gaining or losing electrons, which are effectively weightless. The equivalent weight is simply the ionic weight divided by the charge.

Atoms, molecules and ions are very small: it takes 6.02×10^{23} hydrogen atoms to form a gram. This number is known as *Avogadro's number*: if the weights of the individual species listed in Tables 4.1 and 4.2 are multiplied by this number then the numbers listed in the second and third columns become the weights in grams. They are then referred to generally as the *molar* weights. Dividing the concentration in g/L of any substance by the molar weight then yields the *molar concentration* of the substance, and dividing by the equivalent weight yields the *equivalent concentration*. These parameters become significant when determining quantities of chemical reagents required to carry out chemical reactions (Section 4.5).

4.3 SALTS, ACIDS AND BASES

Ions of opposite charge cling together by electrical attraction (or *ionic bonding*) to make compounds called *salts*. Salts which are or can be dissolved to form a solution are sometimes referred to as *electrolytes*: common salt NaCl is made of Na^+ and Cl^- ions and dissolves (or *dissociates*) readily in water to form these free ions (Equation 4.1). The total charge of a salt or electrolyte must be zero for electrical neutrality to be preserved. So, for example, two Na^+ ions would combine with a single sulphate ion SO_4^{2-} to form the salt Na_2SO_4 (sodium sulphate) which dissolves/dissociates in water to form free ions (Equation 4.2). Salts make up most of the material dissolved in natural and potable waters; wastewaters tend to contain more organic matter whose chemistry is more complicated.

$$NaCl \Rightarrow Na^+ + Cl^- \tag{4.1}$$

$$Na_2SO_4 \Rightarrow 2Na^+ + SO_4^{2-}. \tag{4.2}$$

A substance which dissociates in water forming hydrogen ions as the only cation is called an *acid*. The most common acids used in water and wastewater treatment include hydrochloric acid (HCl) and sulphuric acid (H_2SO_4), both of which fully dissociate in water according to Equations 4.3 and 4.4 respectively.

$$HCl \Rightarrow H^+ + Cl^- \tag{4.3}$$

$$H_2SO_4 \Rightarrow 2H^+ + SO_4^{2-}. \tag{4.4}$$

A substance which dissociates in water forming an anion which can quench the acid ion by chemical reaction is called a *base* or an *alkali*. These usually contain hydroxide ions (OH^-) as the anion, examples being sodium and calcium hydroxide:

$$NaOH \Rightarrow Na^+ + OH^- \tag{4.5}$$

$$Ca(OH)_2 \Rightarrow Ca^{2+} + 2OH^-. \tag{4.6}$$

Another key example is sodium bicarbonate ($NaHCO_3$), which dissociates to produce bicarbonate (HCO_3^-) which then reacts with acid to produce carbon dioxide (CO_2).

When an acid reacts with an alkali the products of the reaction are a salt plus water:

$$HCl + NaOH \Rightarrow NaCl + H_2O. \tag{4.7}$$

Equation 4.7 demonstrates the key chemical reaction of quenching of acids or bases by combing the two to produce a *neutral* solution (see Section 4.5.2.1). Given that both HCl and NaOH completely dissociate in water (according to Equations 4.3 and 4.5 respectively), then if just the acid and base ions (H^+ and OH^- respectively) are considered the equation becomes:

$$H^+ + OH^- \Rightarrow H_2O. \tag{4.8}$$

This is a key chemical reaction in water treatment, since it results in the adjustment of the *pH* (or acidity) of the water.

4.4 CONCENTRATION AND IONIC BALANCE

Because chemicals in water react on an atomic, ionic or molecular basis, their concentrations in mass per unit volume, normally milligrammes per litre (mg/L, sometimes expressed as *parts per million* or *ppm*), must be converted to one on a chemical basis by dividing by the atomic, ionic or molecular weight. This then produces the molar concentration in mmoles/L (or *mM*). Molar concentration is sometimes depicted in square brackets: e.g. "$[O_2]$" indicates the concentration of oxygen and "$[OH^-]$" the concentration of hydroxyl ions in moles per litre (or *M*, which is 1000 mM). According to the data in Table 4.1, 10 mg/L of hydrochloric acid has a molar concentration of $10/(1 + 35.5) = 0.274$ mM (and so $[HCl] = 0.000274$ M). 10 mg/L of NaOH, on the other hand, equates to $10/(23 + 17) = 0.25$ mM (or 0.00025 M, if expressed as $[NaOH]$).

Equivalent concentrations provide the concentration in terms of charge. For a solution to be electrically neutral, the total charge provided by the cations must equal that of the anions. So, for any solution of ions in water, the sum of the equivalent concentrations of cations must equal that of the anions. For example, a water having the specification below in Table 4.3 has almost three times the mass concentration in mg/L of anions than of cations, but the equivalent concentrations are the same and the analysis is therefore valid.

Since ions in water are charged, they can carry electricity and so have an electrical *conductivity*. Electrical conductivity provides a quick and simple indication of the total dissolved solids (TDS) concentration in the water. Conductivity is given in units of µS/cm, S being the SI unit, the *siemens*, and equates to a reciprocal ohm (i.e. $1\,S = 1\,\Omega^{-1}$). A **very** crude approximation relating conductivity to concentration is:

$$1 \text{ meq/L of ions } \sim 100 \text{ µS/cm conductivity.} \tag{4.9}$$

This relationship breaks down at high or low pH levels, since it assumes that all ions move at the same speed when exposed to the same electromotive force. This is a reasonable assumption for all ions other than the hydrogen and hydroxide ions which, because of their chemical nature, are able to move much more rapidly. It also breaks down at high ionic concentrations when the ions start to associate (i.e. re-form salts or molecules).

EXAMPLE: MOLAR CONCENTRATION FROM MASS

What is the molar concentration of a 5 mg/L solution of sulphuric acid?

Molar concentration = mass concentration/molar weight

Substituting the atomic weight values from Table 4.1

Molar concentration = $5/(2 \times 1 + 32 + 4 \times 16) = 5/98 = 0.051$ mM H_2SO_4

Table 4.3 Typical ground water analysis.

Cation	Concentration		Anion	Concentration	
	mg/L	meq/L		mg/L	meq/L
Ca^{2+}	76	3.8	HCO_3^-	244	4.0
Mg^{2+}	14	1.2	Cl^-	38	1.1
Na^+	35	1.5	SO_4^{2-}	52	1.1
			NO_3^-	21	0.3
TOTAL	125	6.5		355	6.5

EXAMPLE: MASS CONCENTRATION FROM WATER SPECIFICATION

What must be the mass concentration of sodium be if a solution containing only sodium, chloride, calcium and bicarbonate ions contains 350 mg/L chloride, 40 mg/L calcium and 100 mg/L bicarbonate?

Total anion equivalent concentration (Table 4.1) $= 350/35.5 + 100/61$

$$= 9.86 + 1.64 = 11.5 \text{ meq/L.}$$

Equivalent concentration of calcium (Table 4.2) $= 40/20 = 2$ meq/L
So, equivalent concentration of sodium $= 11.5 - 2 = 9.5$ meq/L
Therefore, mass concentration of sodium $= 9.5 \times 23 = 219$ mg/L

EXERCISE 4.1
What is the mass concentration of a 0.15 M solution of sodium sulphate?

EXERCISE 4.2
A laboratory reports the following water analysis all in mg/L

Calcium, Ca^{2+}	*118*
Magnesium, Mg^{2+}	*5*
Bicarbonate, HCO_3^-	*280*
Chloride, Cl^-	*57*
Sulphate, SO_4^{2-}	*96*

What concentration of sodium in mg/L gives a balanced analysis?

Concentration may also be expressed with reference to a specific element. Most commonly, the concentrations of nitrogen-containing ions like nitrite, nitrate, and ammonia are often expressed "as N". A 1 mM solution of NO_3^- contains 62 mg/L as NO_3^-, and 1 mM of NH_3 contains 17 mg/L as NH_3, but both could equally be written as ""14 mg/L as N"". Wastewaters contain a lot of nitrogenous compounds which change by biochemical or chemical reaction from one species to another, so it is convenient to know the total content of nitrogen. This demands the converting of all the individual measurements to "mg/L as N" before adding them together. Much the same calculation is carried out with the impurities which cause hardness in water (primarily calcium and magnesium), and also carbonate/bicarbonate. A somewhat dated way of expressing equivalent concentration is as "mg/L (or ppm) as $CaCO_3$". This is simply the molar concentration multiplied by 100 (i.e. the molar mass of $CaCO_3$), or the equivalent concentration multiplied by 50.

Lastly, concentrations may be expressed logarithmically, a very common shorthand in particular for hydrogen ion concentration. The terminology used is p, which represents the negative log of the concentration in moles per litre. Hence, in the case of the hydrogen ion, H^+ (Section 2.3):

$$pH = -\log_{10}[H^+] \tag{4.10}$$

Similarly, $pOH = -\log_{10}[OH^-]$ and $pCa = -\log_{10}[Ca^{2+}]$.

EXERCISE 4.3
If a water sample has a pCa value of 2.5. What is the concentration of calcium in mg/L? If a further 4 mM of Ca is added to the solution, what is the new pCa?

Figure 4.1 offers a quick reference for conversion of concentration units between ppm, mM, meq/L, ppm "as $CaCO_3$" and conductivity in μS/cm. The conversion to conductivity involves the use of an empirically derived proportionality coefficient,

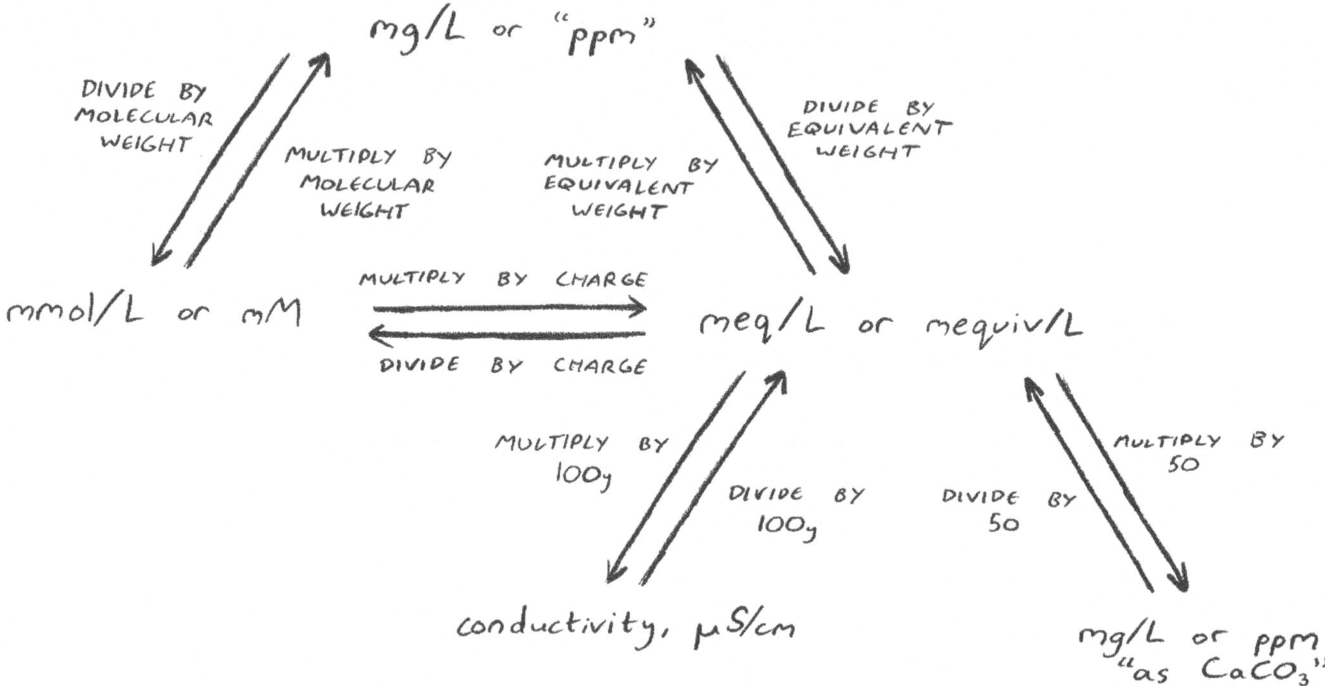

Figure 4.1 Conversion of ionic concentrations.

denoted y. This factor accounts for the *non-ideality* of real solutions, with increasing deviation from ideal solution behaviour (on which Equation 4.9 is based) with increasing concentration.

4.5 STOICHIOMETRY
4.5.1 Stoichiometry and balancing equations

The *stoichiometry* of a chemical reaction relates to the molar ratio with which the chemicals react and the products are formed, based on the assumption that the reaction goes to completion (i.e. that all the reactants are completely converted to products). Whilst something of a simplification, this assumption allows the calculation of the amounts of reagents needed to carry out chemical processes. Since all chemical reactions proceed on a molecular level the relative amounts of chemicals involved can be calculated from the ratio of the number of molecules of each chemical species involved in the governing chemical equation. So, for example, in Equation 4.7 the stoichiometric ratio of the HCl, NaOH and NaCl is 1:1:1. This means that the corresponding weight ratio, according to Tables 4.1 and 4.2, is 36.5:40:58.5. On the other hand, for the precipitation reaction (where formation of a solid is denoted "↓"):

$$AlCl_3 + 3H_2O \Rightarrow Al(OH)_3 \downarrow + 3HCl \tag{4.11}$$

the stoichiometric ratio of hydroxide ions (OH^-) to aluminium ions (Al^{3+}) in aluminium hydroxide is 3:1. This equation implies that dosing of water with aluminium salts influences the solution pH because HCl is generated by *hydrolysis*, i.e. reaction with water.

EXAMPLE: PRECIPITATION OF ALUMINIUM HYDROXIDE

What mass of hydrochloric acid is generated by 2 kg of aluminium chloride and what mass of aluminium hydroxide is precipitated?
From Equation 4.11, 1 mole of $AlCl_3$ generates 3 moles of HCl and 1 mole of $Al(OH)_3$.

From Table 4.1, the molar weights are:

$AlCl_3 = 27 + (3 \times 35.5) = 133.5$ g

$HCl = (35.5 + 1) = 36.5$ g

$Al(OH)_3 = 27 + 3 \times (16 + 1) = 78$ g

So, for the reaction stoichiometry of 3 moles HCl to one mole $AlCl_3$ and $Al(OH)_3$:

133.5 g $AlCl_3$ hydrolyses to produce 78 g $Al(OH)_3$ and $3 \times 36.5 = 109.5$ g HCl

So, 2 kg $AlCl_3$ yields $2 \times 78/133.5 = 1.17$ kg $Al(OH)_3$ and $2 \times 109.5/133.5 = 1.64$ kg HCl

A fundamental principle of chemistry is that the products and reactants of a chemical equation are in perfect balance: there is no net gain of either material or electrical charge. The balance in charge arises from the *redox* principle, which states that oxidation of one species of a chemical reaction must necessarily give rise to reduction of another. For example, the oxidation of Fe^{2+} can be expressed in many forms – all of which obey the fundamental rule of material and charge balance:

$$Fe^{2+} \Rightarrow Fe^{3+} + 1e^- \tag{4.12}$$

$$4Fe^{2+} + O_2 + 10H_2O \Rightarrow 4Fe(OH)_3\downarrow + 8H^+ \tag{4.13}$$

$$2Fe^{2+} + 1/2\,O_2 + 4OH^- + H_2O \Rightarrow 2Fe(OH)_3\downarrow \tag{4.14}$$

In the first example (Equation 4.12) the reaction is expressed simply as a transfer of an electron ("$1e^-$") from the ferrous ion Fe^{2+} to an unidentified source. The second and third equations identify the source as the oxygen molecule, which is then reduced to the hydroxide ion (OH^-) and combines with the ferric ion (Fe^{3+}) to form ferric hydroxide ($Fe(OH)_3$). Whilst the latter two equations may look different, they are both valid expressions of the same redox reaction and have the same 4:1 $Fe^{2+}: O_2$ stoichiometry.

EXAMPLE: PRECIPITATION OF FERRIC HYDROXIDE

How much ferric hydroxide is produced when 50 litres of 25 mg/L ferrous solution is fully oxidised?

From Table 4.1, molar weight of Fe = 56. So, 25 mg/L $Fe^{2+} = 25/56$ mM $= 0.446$ mM Fe
MW ferric hydroxide $Fe(OH)_3 = 56 + 3 \times (16 + 1) = 107$.

Stoichiometry of $Fe(OH)_3$ to Fe = 1:1, according to Equation 4.13.
So, 0.446 mM $Fe(OH)_3) = 0.446 \times 107 = 47.8$ mg/L
and for 50 L the mass of $Fe(OH)_3$ generated is $50 \times 47.8 = 2390$ mg, or 2.39 g $Fe(OH)_3$.

The rules governing stoichiometry are not always straightforward. However, it is convenient to regard all elements in compounds as having an *oxidation state* or *oxidation number* (*ON*), which can be the charge on an ion or the valency of a covalently bonded element. For example, iron can have an *ON* of zero in its elemental form, $+2$ in its ferrous form, or $+3$ in its ferric form. Carbon almost always has an *ON* of $+4$ in the organic and inorganic compounds it forms. Chlorine usually takes an *ON* of -1, for example in hydrochloric acid (HCl) and tetrachloroethane (CCl_4), but can also assume an *ON* of $+1$, for example in hypochlorite (OCl^-), and $+7$, for example in perchloric acid ($HClO_4$). Finally, nitrogen can assume oxidation states of -3 and $+5$ in its most stable forms (in ammoniacal substances and nitrate respectively) but can also assume many other oxidation states in more reactive substances, such as nitrous oxide (N_2O, $ON = +1$), nitric oxide (NO, $ON = +2$), nitrite (NO_2^-, $ON = +3$) and nitrogen dioxide (NO_2, $ON = +4$). The oxidation number of metals is often indicated as a Roman numeral postscript, for example Fe(III) or Cr(VI).

For most chemical reactions encountered in water treatment the following six basic rules can usually be applied to determine the stoichiometry and overall chemical reaction:

(1) There can be no net change in the quantity of each element.
(2) There can be no net gain or loss of charge. Thus, in redox reactions: (a) the sum of the ionic charges on one side of the equation must be the same as that on the other, and (b) the sum of the total oxidation numbers on either side of the equation must also be the same.

(3) Hydrogen takes a value of $+1$ unless it is either (a) in the elemental state as H_2, in which case it is zero or, very rarely, (b) in the hydride form (H^-), in which case it is -1.

(4) Oxygen takes a value of -2 unless it is either (a) in the elemental state as O_2 or O_3, when it is zero, or (b) in the *peroxo* state (for example in hydrogen peroxide H_2O_2) in which case it is -1.

(5) In their most stable forms:
  - *Alkaline* or *Group I* metals (Li, Na, K, Rb and Cs) take an O.N. value of $+1$,
  - *Alkaline Earths* or *Group II* metals (Mg, Ca, Sr, Ba) take an O.N. value of $+2$.
  - *Halogens* (F, Cl, Br and I) usually take an *ON* of -1 or, if attached to a nitrogen or oxygen atom, a value of $+1$,

(6) If there is an imbalance of material, it can only be made up with water or its products H^+ or OH^-.

Balancing of any chemical equation proceeds in three stages:

(1) **Oxidation number calculation**
     ON values normally calculated from the ion charge and/or the assumption of values of $+1$ and -2 for combined hydrogen and oxygen respectively.

(2) **Charge balance**
     Ratio of the reactants and products from the redox principle.

(3) **Material balance**

     Quantity of additional atoms (O and H) required for equation to balance.

EXAMPLE: OXIDATION WITH PERMANGANATE

Dissolved iron (Fe^{2+}) reacts with permanganate (MnO_4^-) to produce the solids ferric floc ($Fe(OH)_3$) and MnO_2 – the most stable form of manganese. Thus:

Reactants: $Fe^{2+} + MnO_4^-$
Products: $Fe(OH)_3 + MnO_2$

Write a balanced equation for this reaction

There are three steps to the development of the equation:

(1) *Oxidation number calculation*
     Assuming appropriate *ON* values for H ($+1$) and O (-2), the *ON* values for the reacting species are:

     Fe: From $+2$ in the reactant to $+3$ in the product, a change of $+1$
     Mn: From $+7$ in the reactant to $+4$ in the product, a change of -3

     Hence for redox balance, the electrons taken up by the Mn must be provided by the Fe. Thus, 1 mole of Mn reacts with 3 moles of Fe. It follows that there must also be three moles of $Fe(OH)_3$ product:

$$3Fe^{2+} + MnO_4^- \Rightarrow 3Fe(OH)_3 + MnO_2$$

(2) *Charge balance*
     The sum of the positive and negative charges on either side of the above equation are:

     Reactants: $(3x + 2) + (1x - 1) = +5$
     Products: zero

     Hence for charge balance, either:
       (a) 5 moles of negative charge must be added to the reactants (assumed to be hydroxide ion, OH^-) or
       (b) 5 moles positive charge must be added to the products (assumed to be H^+).
It doesn't matter which of the assumptions, (a) or (b), is made: both will give the correct answer.

(3) *Material balance*
     Assuming five hydroxide ions are added to the left hand side:

$$3Fe^{2+} + MnO_4^- + 5OH^- \Rightarrow 3Fe(OH)_3\downarrow + MnO_2\downarrow$$

Since Fe and Mn both balance, only the oxygen and hydrogen are not balanced, in that:

- For O, there is a 2 mole deficit on the reactant side, and
- For H, there is a 4 mole deficit on the reactant side.

Thus, there are two H_2O molecules needed on the left-hand side (LHS) of the equation for it to balance, and the final balanced equation is thus:

$$3Fe^{2+} + MnO_4^- + 5OH^- + 2H_2O \Rightarrow 3Fe(OH)_3\downarrow + MnO_2\downarrow$$

Note that if, for the charge balance, $5H^+$ ions are added to the RHS of the equation then the final equation becomes:

$$3Fe^{2+} + MnO_4^- + 7H_2O \Rightarrow 3Fe(OH)_3\downarrow + MnO_2\downarrow + 5H^+$$

This equation is as representative of the reaction as the one above it, the only difference being that 5 H^+ ions have been added to either side.

EXERCISE 4.4

The ammonium ion (NH_4^+) can be fully oxidised to nitrate (NH_3^-) by hypochlorous acid (HOCl), which is reduced to chloride in the process. Write a balanced equation for this reaction.

4.5.2 Key chemical reactions in water and wastewater treatment

There are a large number of chemical/biochemical reactions which can take place in water and wastewater treatment, and the following are a few examples. Whilst stoichiometric considerations allow the ratio of a chemical reagent to a contaminant to be determined, it is normal for a *stoichiometric excess* to be employed in practice to ensure that the reaction goes to completion.

4.5.2.1 pH adjustment and neutralisation

An important chemical process in water and wastewater treatment is the adjustment of pH where, according to Equation 4.10:

$$pH = -\log_{10}[H^+]$$

According to equilibrium thermodynamics (see Section 4.6):

$$pH + pOH = 14 \tag{4.15}$$

This means that at the lowest concentrations of both H^+ and OH^-, $pH = pOH = 7$ (i.e. $[H^+] = [OH^-] = 10^{-7}$). Under these conditions, pH 7, the water is considered to be *neutral* with respect to acidity and alkalinity. Thus, *neutralisation* of a solution, i.e. adding the same molar concentration of acid as there is base or vice versa, produces a solution containing 10^{-7} M each of H^+ and OH^-. The logarithmic relationship means the impact on pH of adding acid or base depends on the pH of the solution. For example, the addition of 10^{-6} moles per litre of acid or hydroxide to a neutral solution (pH 7) changes the pH by one unit. If the same dose is added to a solution at pH 4 the pH barely changes, since the acid concentration at pH 4 is 1000 times that at pH 7 and a 100 times greater than the added dose. This is a consequence of the water equilibrium (Section 4.6.3).

4.5.2.2 Iron oxidation

Many borehole waters contain quantities of the dissolved ferrous ion (Fe^{2+}, 56 g/mol according to Table 4.2), which can be oxidised to produce ferric hydroxide precipitate ($Fe(OH)_3$, $56 + 3 \times (16 + 1) = 107$ g/mol) using air (Equations 4.12–4.14). Other more powerful oxidants, such as permanganate (MnO_4^-, $55 + (4 \times 16) = 119$ g/mol), may also be used to achieve this conversion more rapidly (see previous example):

$$3Fe^{2+} + MnO_4^- + 5OH^- + 2H_2O \Rightarrow 3Fe(OH)_3\downarrow + MnO_2\downarrow \tag{4.16}$$

According to the above equation, 119 g MnO_4^- oxidizes $3 \times 56 = 168$ g/mol g Fe^{2+}. The reaction is base catalysed; hydroxide (OH^-) appears the left-hand side of the equation. If the reaction can be assumed to go to completion then 0.71 mg MnO_4^- of permanganate is required to oxidise 1 mg of iron, provided there are no side reactions which might deplete the permanganate. In practice, permanganate is rarely used for this duty, partly because of its tendency to produce a pink colouration in the product water.

Many toxic metals can be removed by precipitation through pH adjustment since, as with iron, they form sparingly soluble hydroxide precipitates which can then be removed by gravitation or filtration. The solubility of these metals is thus pH dependent, and governed by the *solubility product* (see Section 4.6.2) of the hydroxide salt.

4.5.2.3 Carbonate precipitation

Lime ($Ca(OH)_2$) is sometimes used to soften water, i.e. reduce levels of hardness (Ca^{2+} and Mg^{2+} ions) as well as removing alkalinity (HCO_3^- ions), according to the following chemical reaction:

$$Ca(OH)_2 + Ca(HCO_3)_2 \Leftrightarrow 2CaCO_3\downarrow + 2H_2O. \tag{4.17}$$

Thus, 1 mole of calcium bicarbonate (molecular weight 162 g) reacts with 1 mole of lime (molecular weight 74 g) to produce 2 moles of calcium carbonate product (molecular weight 100 g). If the reaction goes to completion, this would imply that 0.46 mg of lime would need to be added for each mg of calcium bicarbonate for all the calcium to be precipitated.

4.5.2.4 Chlorination of ammoniacal solutions

The reaction between chlorine and ammonia can take place via a number of chemical pathways:

$$NH_3 + HOCl \Rightarrow NH_2Cl + H_2O \tag{4.18}$$

$$NH_3 + 2HOCl \Rightarrow NHCl_2 + 2H_2O \tag{4.19}$$

$$2NH_3 + 3HOCl \Rightarrow N_2 + 3H^+ + 3Cl^- + 3H_2O \tag{4.20}$$

$$NH_4^+ + 4HOCl \Rightarrow NO_3^- + 6H^+ + 4Cl^- + H_2O \tag{4.21}$$

The stoichiometric ratio of chlorine to ammonia can thus vary between 1:1 and 4:1, according to the above expressions, depending how far the reaction proceeds (i.e. how oxidised the nitrogen becomes). Complete oxidation is often referred to as *breakpoint chlorination*. Partial oxidation to the chlorine-substituted ammoniacal compounds NH_2Cl and $NHCl_2$, collectively called *chloramines*, is carried out for the disinfection process *chloramination*. Chloramination is sometimes used in place of chlorination for final disinfection of drinking water, since chloramines are more chemically stable (though less powerful as biocides) than dissolved chlorine.

4.5.2.5 Biochemical oxidation

The significance of reaction stoichiometry is not limited solely to chemical reactions, since biochemistry is a crucial part of wastewater treatment. In biological treatment, micro-organisms undergo *endogenous respiration*, i.e. biochemically oxidize part of their own cells instead of new organic matter entering the process, according to the equation:

$$C_5H_7NO_2 + 5O_2 \Rightarrow 5CO_2 + 2H_2O + NH_3 + \text{energy} \tag{4.22}$$

where $C_5H_7NO_2$ represents bacterial cell material. Hence, according to Equation 4.22, 5 moles of molecular oxygen (O_2) are needed to allow 1 mole of cell material to be converted to carbon dioxide, water and ammonia. This equates to a molar weight ratio of 160:113, or 1.42:1, oxygen:cells for complete oxidation. This then relates to the *biochemical oxygen demand* (BOD) of the wastewater, which is therefore 1.42 times the concentration by mass of the biodegradable cellular material if endogenous respiration takes place exclusively.

4.6 EQUILIBRIUM THERMODYNAMICS

For some chemical reactions in water and wastewater treatment there is a significant degree of reversibility, such that both reactants and products exist at significant concentrations. In such cases, the arrow "$\Rightarrow$" must be replaced by a double-arrow "$\rightleftharpoons$". Considering the general *reversible* chemical reaction where reactants A and B form products C and D:

$$a\text{A} + b\text{B} \rightleftharpoons c\text{C} + d\text{D} \tag{4.23}$$

where a, b, c and d represent the number of moles of each component existing at equilibrium. According to the *Law of Mass Action*, when the reaction reaches *equilibrium* (i.e. the forward and reverse reactions proceed at the same rate) the ratio of reactants to products is given by:

$$K = \frac{[\text{C}]^c[\text{D}]^d}{[\text{A}]^a[\text{B}]^b} \tag{4.24}$$

where K is known as the *equilibrium constant* and is constant for any chemical reaction at equilibrium at a given temperature and pressure. So, if the reaction conditions were to be disturbed by, for example, adding more B or removing some of C, the concentration of the other components would readjust to produce the same value of K. Thus, if both K and the reaction stoichiometry is known, then the ratio of the reactants and products can be calculated.

Many examples of dynamic physical/chemical equilibria exist, where reactants constantly and rapidly interchange with products. In all cases the Law of Mass Action is always obeyed, and the balance of reactants and products is therefore defined by some equilibrium constant.

4.6.1 Weak acids and bases

Compounds which completely dissociate in water to their constituent ions are known as *strong electrolytes*. Some electrolytes do not dissociate completely and are defined as *weak*. Common weak acids and bases include acetic acid (vinegar) and ammonium hydroxide respectively. Dissociation can be regarded as a reversible equilibrium reaction with, in the case of an acid, an *acid dissociation constant*, K_a. For the dissociation of acetic acid:

$$\text{HAc} \rightleftharpoons \text{Ac}^- + \text{H}^+ \tag{4.25}$$

where Ac^- is the acetate ion CH_3COO^-, K_a is defined as:

$$K_a = \frac{[\text{H}^+][\text{Ac}^-]}{[\text{HAc}]} \tag{4.26}$$

Since dissociation constants of weak acids are usually very small, it is convenient to express them in the logarithmic p form, where $pK_a = -\log_{10}K_a$ and decreases with increasing acid strength. Equation 4.26 implies that when the pH value equals pK_a, the concentrations of HA and Ac^- are equal (i.e. dissociation is half complete). At lower pH values (more acidic solutions) dissociation is reduced and at higher values it is increased: when the pH is greater than pK_a by more than two units, dissociation is >99% complete.

For weakly basic substances, such as ammonia, dissociation increases at low pH levels to generate hydroxide along with the ammonium ion:

$$\text{NH}_3 + \text{H}_2\text{O} \rightleftharpoons \text{NH}_4\text{OH} \rightleftharpoons \text{NH}_4^+ + \text{OH}^- \tag{4.27}$$

In this case the base dissociation constant K_b can be defined:

$$K_b = \frac{[\text{NH}_4^+][\text{OH}^-]}{[\text{NH}_3]} \tag{4.28}$$

where the concentration of water in the denominator can be ignored. Equation 4.27 can also be written as:

$$\text{NH}_4^+ \rightleftharpoons \text{NH}_3 + \text{H}^+ \tag{4.29}$$

with a corresponding acid dissociation constant K_a (of around 9.3), defined as in Equation 4.26 with $[NH_3]$ and $[NH_4^+]$ respectively replacing $[Ac^-]$ and $[HAc]$.

EXAMPLE: AMMONIA EQUILIBRIUM

How much free ammonia (NH_3) is (a) initially present, and (b) produced by the addition of 40 g of NaOH to 1 tonne of water containing 300 mg/L "as N" of dissolved ammonia and having a pH of 3? Assume an acid dissociation constant of 9.3.

From Equation 4.29, manipulating the logarithms according to Table 2.2 and substituting "p" for "−log":

$$pK_a = 9.3 = \log\frac{[NH_3][H^+]}{[NH_4^+]} = pNH_3 + pH - pNH_4^+ \tag{A}$$

The total ammonia concentration = 300 mg/L "as N"; the molar concentration is obtained by dividing by the atomic weight of nitrogen. Thus:

$$[NH_3] + [NH_4^+] = 300 \times 10^{-3}/14 = 21.4 \times 10^{-3}.$$

It can be seen that the pH at which $[NH_3] = [NH_4^+]$ is 9.3. For each displacement of 1 pH unit away from this value $[[NH_3]/[NH_4^+]$ decreases by an order of magnitude. Hence, at pH 3, $[NH_3]$ is more than five orders of magnitude ($>$100,000) less than $[NH_4^+]$.

So, assuming all ammonia is as NH_4^+ at pH 3, the initial $[NH_4^+] = 21.4 \times 10^{-3}$, and so $pNH_4^+ = -\log(21.4 \times 10^{-3}) = 1.67$. From Equation A above:

Initial pNH_3 at pH 3 $= 9.3 - 1.67 - 3 = 7.97$

Initial $[NH_3] = 10^{-7.97} = 1.07 \times 10^{-8}$ M, or 0.15 µg/L as N.

If 40 g NaOH is added, moles OH^- added $= 40/40 = 1$.

1 mole per tonne water $= 1/1000$ moles/litre, and thus $[OH^-]$ added $= 10 - 3$ M; pOH $= 3$

So, $[OH^-]$ added $= [H^+]$ present; the water is thus neutralised and the new pH is 7.

Thus, from Equation A again:

New pNH_3 at pH 7 $= 9.3 + 1.67 - 7 = 3.97$

New $[NH_3] = 1.07 \times 10^{-4}$ M, or 1.5 mg/L as N.

EXERCISE 4.5

A 50 mg/L solution of formic acid (HFo, Fo^- having the formula CHO_2) at pH 4 is to be dosed with 50 mg/L HCl. What is the concentration of formate (Fo^-) concentration before and after dosing if the pKa of formic acid is 3.75?

4.6.2 Dissolution of salts

As with acids and bases, many salts do not completely dissociate in water but are weak electrolytes with limited solubility. Common examples in natural waters include salts of carbonate (CO_3^{2-}, Section 4.6.5), and silicate ($HSiO_3^-$), both of which combine with divalent cations to form sparingly soluble *scales*. The solubility of most solids increases with temperature, although there are some important exceptions to this. At the limit of its equilibrium solubility, a compound is said to form a *saturated* solution in the water. If the ions reach a concentration higher than saturation, they precipitate – as may happen if a saturated solution is cooled.

For a solid salt M_mA_n dissolution is represented by the equilibrium:

$$M_mA_n \rightleftharpoons mM^{n+} + nA^{m-} \tag{4.30}$$

where M is the cation and A the anion. The equilibrium constant in this case is referred to as the *solubility product*, and is given by:

$$K_{SP} = [M]^m[A]^n$$

Table 4.4 Salt solubility values at 20°C.

Salt	Solubility mg/L	Solubility product K_{SP}
NaCl	263,000	1900
NaHCO$_3$	65,000	9.3
CaSO$_4$	2980	1.9×10^{-4}
CaF$_2$	8	3.2×10^{-11}
CaCO$_3$	1.2	8.7×10^{-9}

or

$$\log K_{SP} = m \log[M] + n \log [A] \tag{4.31}$$

This is analogous to the dissociation equation in the previous example, but in the above case the concentration of the reactant in the water $[M_mA_n]$ is ignored. It follows that if the product of the terms on the RHS of Equation 4.31 exceeds the K_{SP}, precipitation takes place. So, for example, if sodium sulphate is added to a solution of calcium chloride then the solubility product of calcium sulphate may be exceeded, and this salt precipitated as a result. It is also possible for the concentration of ions to exceed the natural solubility dictated by K_{SP} and yet remain dissolved, in which case the solution is described as *supersaturated*.

Values for the equilibrium solubility of some salts at ambient temperatures in pure water are given in Table 4.4. Note that the units of K_{SP} vary according to the number of ions in the salt molecule.

EXAMPLE: CALCIUM SULPHATE DISSOLUTION

How much calcium will dissolve in water containing 100 mg/L sulphate?

The K_{SP} of CaSO$_4$ is 1.9×10^{-4}, according to Table 16.

The water contains 100 mg/L sulphate, i.e. $100 \times 10^{-3}/96 = 1.04 \times 10^{-3}$ M (for a molar mass of 96 g, Table 4.2). From Equations 4.30 and 4.31:

$$K_{SP} = [Ca^{2+}][SO_4{}^{2-}] = [Ca^{2+}]1.04 \times 10^{-3}$$

So $[Ca^{2+}] = 1.9 \times 10^{-4}/1.04 \times 10^{-3} = 0.182$ M, i.e. 7.3 g/L at 40 g Ca/mol (Table 13).

EXERCISE 4.6
A water containing 120 mg/L Ca^{2+} is saturated with fluoride. If no other anions are present, what is the concentration in mg/L of fluoride?

EXERCISE 4.7
Assuming the K_{sp} of lead hydroxide, Pb(OH)$_2$, to be 4×10^{-15}, what pH is required to achieve a dissolved lead concentration of 0.1 μg/L if the Pb solution concentration is 0.15 mg/L at pH 7? The molar mass of Pb is 207 g/mol.

4.6.3 Dissociation of water and pH

The dissociation of water (Equation 4.8) and its relationship to pH has been defined in Equations 4.10 and 4.15:

$$pH = -\log_{10}[H^+]$$

and $pH + pOH = 14$.

Water thus behaves as a very weak acid (like acetic acid in Equation 4.25) and has an acid dissociation constant of 10^{-14}, normally denoted K_w, which is the product of the molar acid and hydroxide concentrations; it follows that $pK_w = 14$. As stated in

Section 4.5.2.1, water always contains acid and hydroxide ions – even "pure" water in the neutral state (pH $= 7$), where these concentrations are both at 10^{-7} M, or 0.1 μM.

EXAMPLE: ALKALINE PH

How much hydroxide does a solution at pH 10.5 contain?

pOH $= 14 -$ pH $= 3.5$

Thus, $[OH^-] = 10^{-3.5} = 3.16 \times 10^{-4}$ M, or 5.3 mg/L (for a molar mass of 17 g, Table 13).

A key aspect of water treatment is pH adjustment, and there are four scenarios of pH adjustment of *unbuffered* solutions:

(a)   Lowering the pH of an acid solution (i.e. water containing excess H^+, protons)
(b)   Raising the pH of an acid solution
(c)   Raising the pH of an alkaline solution (i.e. water with excess OH^-, hydroxide)
(d)   Lowering the pH of an alkaline solution

If the solution contains bicarbonate (HCO_3^-) a different chemistry presides (Section 4.6.5.1). Bicarbonate is a *buffer*: it tends to resist a change in pH (Section 4.6.5.2).

4.6.3.1 Lowering the pH of an acid solution

If the pH of a solution is less than 7 it is acidic. Adding acid lowers the pH and so simply adds to the molar concentration of acid:

$$pH_{new} = -\log(10^{-existing\ pH} + [H^+]_{added}). \tag{4.32}$$

4.6.3.2 Raising the pH of an acid solution

If the pH of a solution is less than 7 and alkaline is added to raise the pH, then the hydroxide (OH^-) mops up the acid (H^+) to make water. If less hydroxide than there is acid is added then:

$$pH_{new} = -\log(10^{-existing\ pH} - [OH^-]_{added}) \tag{4.33}$$

where $10^{-existing\ pH}$ is simply the existing H^+ molar concentration.

If more alkaline than there is acid is added, then an excess of hydroxide results:

$$pOH_{new} = -\log([OH^-]_{added} - 10^{-existing\ pH}).$$

Thus the new pH is:

$$pH_{new} = 14 - pOH = 14 + \log([OH^-]_{added} - 10^{-existing\ pH}). \tag{4.34}$$

EXERCISE 4.8
If the pH of the solution is 3.5 and a 0.0025 M of hydroxide is added, what would the new pH be?

4.6.3.3 Raising the pH of an alkaline solution

If the pH of a solution is more than 7 it is alkaline (or basic). Adding hydroxide raises the pH and so simply adds to the molar concentration of hydroxide:

$$pOH_{new} = -\log(10^{-existing\ pOH} + [OH^-]_{added})$$

where existing pOH $= 14 -$ existing pH, and thus the new pH is:

$$pH_{new} = 14 + \log(10^{-existing\ pOH} + [OH^-]_{added}) \tag{4.35}$$

EXAMPLE: ALKALINE PH ADJUSTMENT UPWARDS

If the pH of the solution is 10.5 and 0.5 mM of hydroxide is added, what is the new pH?

Since $pOH = -\log[OH]$ and $pOH + pH = 14$ (Equation 4.15), then $[OH^-] = 10^{-(14-pH)}$
So, new $[OH^-] = 10^{-(14-10.5)} + 0.5 \times 10^{-3} = 3.16 \times 10^{-4} + 0.5 \times 10^{-3} = 8.16 \times 10^{-4}$
And new $pH = 14 - pOH = 14 - (-\log(8.16 \times 10^{-4})) = 14 - 3.1 = 10.9$

4.6.3.4 Lowering the pH of an alkaline solution

If the pH of a solution is more than 7 it is basic (or alkaline) and if acid is added to lower the pH, then the protons mop up the hydroxide to produce water. If less acid is added than hydroxide then:

$$pOH_{new} = -\log\left(10^{-\text{existing pOH}} - [H^+]_{added}\right). \tag{4.36}$$

The calculation thus proceeds along the same lines as adding alkali to an acid solution, only with reference to pOH rather than pH.

If more acid than hydroxide is added, then the calculation proceeds in an analogous way to that of adding excess hydroxide to acid, this time with reference to acid (pH) rather than alkali (pOH):

$$pH_{new} = -\log\left([H^+]_{added} - 10^{-\text{existing pOH}}\right). \tag{4.37}$$

EXERCISE 4.9
If the pH of the solution is 12.5 and a 0.04 M of acid is added, what would the new pH be? What would be the new pH of only half this amount of acid was added?

If acid or alkaline reagents are not added as highly concentrated solutions but instead add to the system or flow volume, then it is necessary to complete a mass balance (Chapter 6) to calculate the impact on pH. This may happen when two water sources are blended.

4.6.4 Dissolution of gases

In a mixture of gases each gas exerts its own partial pressure independently of the others, and the total gas pressure is the sum of all the partial pressures. For example, air contains roughly 20% oxygen and 80% nitrogen, and so at 1 bar pressure (i.e. atmospheric) the partial pressure of each is 0.2 bar and 0.8 bar respectively. In contact with water an equilibrium is set up between the gases in the atmosphere and those dissolved in the water. If no chemical reaction takes place, then the ratio of concentration of gas i in the water (c_i) to its partial pressure in the atmosphere (p_i) is given by:

$$p_i = H_i c_i \tag{4.38}$$

where H_i is *Henry's constant* for species i, and is constant for a given gas and temperature and takes units dependent on those of the aqueous concentration (Table 4.5).

If both the gaseous and aqueous concentrations of gas are expressed as mole fractions, then the Henry constant takes units of pressure, such as atmospheres (atm), and the values in column 1 of the table apply. However, it is more usual for the liquid concentration to be given in in mg/L (column 2) or mM (column 3).

EXAMPLE: HENRY'S LAW

Water contaminated with 12 µg/L chloroform is held in an enclosed space. What is the partial pressure of chloroform in the atmosphere?

Henry constant for chloroform (Table 17) $H_{CHCl_3} = 2.55 \times 10^{-5}$ L · atm/mg
$CHCl_3$ concentration in solution $= 12$ µg/L $= 12 \times 10^{-3}$ mg/L

So, according to Equation 4.38:

$$p_{CHCl_3} = H_{CHCl_3} c_{CHCl_3} = 2.55 \times 10^{-5} \times 12 \times 10^{-3} = 3.06 \times 10^{-7} \text{ atm}.$$

Table 4.5 Values for Henry's constant, 20°C.

	H atm	H atm · L/mg	H atm · L/mmol
Air	7.71×10^4	0.046	1.39
Ammonia	0.83	8.8×10^{-6}	1.5×10^{-5}
Bromoform	35	2.5×10^{-6}	6.3×10^{-4}
Carbon dioxide	151	6.2×10^{-5}	2.72×10^{-3}
Chlorine	767	1.95×10^{-4}	0.014
Chloroform	170	2.55×10^{-5}	3.06×10^{-3}
Oxygen	4.3×10^4	0.024	0.77
Ozone	5×10^3	1.87×10^{-3}	0.090

EXERCISE 4.10

Water is saturated with air at 4 atm pressure. The pressure is then reduced to 1 atm, resulting in the release of air bubbles. What quantity of air (in mg/L) is released under these conditions?

4.6.5 Alkalinity

4.6.5.1 Carbonate speciation

The species CO_2, HCO_3^- and CO_3^{2-} are linked by a series of reversible reactions, collectively referred to as *carbonate chemistry* and with specific equilibrium constant values (Table 4.6). When atmospheric air comes into contact with pure water, CO_2 dissolves in water, the amount varying with the partial pressure according to Henry's Law:

$$CO_2(\text{gas}) \rightleftharpoons CO_2(\text{dissolved}) \tag{4.39}$$

Dissolved CO_2 combines with water to form carbonic acid (H_2CO_3), which is extremely unstable and breaks down to form bicarbonate (HCO_3^-) and then, at increasing pH levels, carbonate (CO_3^{2-}):

$$CO_2(\text{dissolved}) + H_2O \rightleftharpoons H_2CO_3 \quad \text{(dissolution)} \tag{4.40}$$

$$H_2CO_3 \rightleftharpoons H^+ + HCO_3^- \qquad \text{(first dissociation)} \tag{4.41}$$

$$HCO_3^- \rightleftharpoons H^+ + CO_3^{2-} \qquad \text{(second dissociation)} \tag{4.42}$$

such that overall:

$$CO_2(\text{dissolved}) + H_2O \rightleftharpoons H_2CO_3 \rightleftharpoons H^+ + HCO_3^- \rightleftharpoons 2H^+ + CO_3^{2-}$$

The instability of carbonic acid is such that its concentration can be ignored. A consideration of the combined equilibria of Equations 4.40 and 4.41 yields:

$$K_{\text{dissolution and 1st dissociation}} = \frac{[H^+][HCO_3^-]}{[CO_2]} \tag{4.43}$$

Table 4.6 Carbonate equilibrium constant values.

Equation	Reaction	Value at 20°C
4.39	Dissolution of CO_2	2.72 atm L/mol
4.43	Dissolution + 1st dissociation	4.14×10^{-7}
4.44	Dissociation of bicarbonate	3.96×10^{-11}

The second dissociation (Equation 4.42) is described by:

$$K_{2\,\text{nd dissociation}} = \frac{[\text{H}^+][\text{CO}_3{}^{2-}]}{[\text{HCO}_3{}^-]} \tag{4.44}$$

EXERCISE 4.11
Ultrapure water is stored under atmospheric air which contains 0.03% CO_2. What is the equilibrium pH?

Since bicarbonate, carbonate and hydroxide ions all react with mineral acids, and are thus "alkaline salts", they all contribute to *alkalinity* in water. The pH does not necessarily have to be above 7 for water to have alkalinity, since $\text{HCO}_3{}^-$ is present down to pH levels of ~4.5. At pH 8.2, it is almost completely converted to carbonate. Alkalinty attributable to hydroxide therefore, refers to pH above values above 8.2. This is known as the *P alkalinity*. The total alkalinity is sometimes referred to as the *M alkalinity*. The relationship between these two parameters and the individual alkalinity ion levels is summarised in Table 4.7.

EXAMPLE: P AND M ALKALINITY

If a water has a P alkalinity of 80 and an M alkalinity of 120, both in mg/L as $CaCO_3$, what is the carbonate concentration?

Since *P alk* > ½ *M alk*, then according to Table 19 the carbonate concentration is:

$c_{carbonate} = 2(\text{M} - \text{P}) = 2 \times (120 - 80) = 80$ mg/L as $CaCO_3$

EXERCISE 4.12
A water analysis reports a water with P alkalinity 120 mg/L as $CaCO_3$ and M alkalinity 290 mg/L as $CaCO_3$. What is the carbonate concentration?

4.6.5.2 Buffering

Equation 4.43 can be rewritten in logarithmic form with the equilibrium constant values from Table 4.3 substituted:

$$\log\frac{[\text{HCO}_3{}^-]}{[\text{CO}_2]} = \text{pH} - 6.38 \tag{4.45}$$

For normal waters in which Ca^{2+} is the major cation the above equation holds between pH 4.5 and 8.2 and can be used to predict changes in pH resulting from acid or alkali dosing. Adding one mole of acid converts one mole of bicarbonate alkalinity to one mole of carbon dioxide, in accordance with Equations 4.41–4.42:

$$\text{HCO}_3{}^- + \text{H}^+ \Rightarrow \text{CO}_2 + \text{H}_2\text{O}$$

Table 4.7 Relationship between *P* and *M* alkalinity and alkaline ion concentrations.

P *Alk*	OH$^-$	CO$_3{}^{2-}$	HCO$_3{}^-$
Nil	Nil	Nil	*M*
<½ *M*	Nil	2*P*	*M* − 2*P*
½ *M*	Nil	*M*	Nil
>½ *M*	2*P* − *M*	2(*M* − *P*)	Nil
M	*M*	nil	Nil

such that dosing of x moles of acid gives:

$$\log \frac{[HCO_3^- - x]}{[CO_2 + x]} = pH - 6.38 \qquad (4.46)$$

Similarly, adding y moles of hydroxide converts CO_2 into bicarbonate:

$$\log \frac{[HCO_3^- + y]}{[CO_2 - y]} = pH - 6.38 \qquad (4.47)$$

The effect of dosing a bicarbonate solution with acid or base, therefore, is to convert carbon dioxide to bicarbonate or vice versa, rather than directly change the concentration of hydrogen ions. The result is only a minor impact on the pH, and this effect is known as *buffering*.

EXAMPLE: ALKALINITY BUFFERING

If a solution at a pH of 7.25 and having a bicarbonate concentration of 12 mM was to be dosed with 1.2 mM of hydroxide, what would be the new pH? What would be the new pH if (a) no alkalinity was present, and (b) if the solution was dosed with acid instead of hydroxide at the same concentration?

$[CO_2]$ of existing solution (Equation 4.45) is given by:

$$[CO_2] = [HCO_3^-]/10^{(pH-6.38)} = 12/10^{(7.25-6.38)} = 1.62 \text{ mM}$$

New pH when $x = 1.2$ mM in Equation 4.47:

$$pH_{new} = \log ([12 + 1.2]/[1.62 - 1.2]) + 6.38 = 7.9$$

This compares to the case when no alkalinity is present (Equation 4.35):

$$pH_{new} = 14 + \log (10^{-(14-7.25)} + 1.2 \times 10^{-3}) = 14 - 2.92 = 11.1$$

If the same bicarbonate solution is dosed with 1.2 mM ($=x$) acid, the new pH becomes:

$$pH_{new} = \log ([12 - 1.2]/[1.62 + 1.2]) + 6.38 = 7.0$$

EXERCISE 4.13
A mains water supply has an alkalinity of 280 mg/L as HCO_3^- and a pH of 8.1. What dose in mg/L of hydrochloric acid (HCl) is needed to reduce the pH to 7.0?

4.6.6 Hardness

Hardness refers to the calcium and magnesium content of waters, which form the sparingly soluble salts. These tend to precipitate as tenacious crystalline deposits in boilers, cooling towers and heat exchangers, and can also form scum in commercial laundries and in food processing and soft drinks manufacturing. Ca^{2+} and Mg^{2+} have similar chemical properties, though in most natural waters the Mg^{2+} concentration is much smaller than that of Ca^{2+}. Water analyses often combine them as "hardness, mg/L as $CaCO_3$". Hardness is normally divided into *temporary* or *carbonate hardness*, which is hardness associated with the bicarbonate ion, and the *permanent* or *non-carbonate* hardness, which is the remaining hardness.

Hardness can be removed by adsorption using an ion exchange process (Section 4.6.8.2) or by reverse osmosis – a process in which a membrane rejects the hardness ions whilst allowing the water through it. An alternative more dated process is the use of lime ($Ca(OH)_2$), which removes the temporary hardness and the associated alkalinity by precipitation of calcium carbonate, as indicated in Equation 4.17:

$$Ca(HCO_3)_2 + Ca(OH)_2 \Rightarrow 2CaCO_3\downarrow + 2H_2O$$

Permanent hardness can be partially removed by the addition of sodium carbonate (soda ash):

$$Ca^{2+} + Na_2CO_3 \Rightarrow CaCO_3\downarrow + 2Na^+ \tag{4.48}$$

Note that neither if these reactions are examples of redox reactions, since the oxidation state of the reactants does not change.

4.6.7 Langelier Saturation Index

The *Langelier Saturation Index* (*LSI*) arises from the combination of logarithmic versions of the carbonate equilibria (Equations 4.43 and 4.44) and the dissociation of calcium carbonate according to the solubility product (K_{SP}, Equation 4.31). This produces an expression for the pH at which the calcium carbonate is at its limit of solubility, denoted pH_S or the *saturation pH*:

$$pH_S = pK_{SP} - pK - pCa - pHCO_3 \tag{4.49}$$

where K is the overall equilibrium constant for the bicarbonate reaction. The difference between the actual pH of the water and its theoretical pH_s value ($pH - pH_s$) is the *LSI index* and indicates whether the water will deposit calcium carbonate (*scaling*) or dissolve it (*corrosive*). Given that K_{SP} and K are nominally constant, pH_s depends only on the calcium and bicarbonate concentrations. In fact these equilibrium constants have some dependency on the dissolved solids concentration and temperature, and the *LSI* is normally either computed or determined from nomograms.

4.6.8 Adsorption

4.6.8.1 Activated carbon

The removal of dissolved organic compounds by activated carbon (AC), either in the granular form (GAC) or powdered (PAC), is characterised by *adsorption isotherms*. The term *adsorption* refers to the material surface rather than its bulk, where the term *absorption* applies. Adsorption isotherm expressions relate the equilibrium concentration of pollutant at the adsorbent surface (q_e, normally in mg pollutant or *adsorbate* per g *adsorbent*) with that in solution (c_e, mg/L) at constant temperature. Two common expressions are the *Langmuir* and *Freundlich* isotherms, with the latter being the most widely used in water and wastewater applications based on AC:

$$q_e = K_F c_e^{1/n} \tag{4.50}$$

where K_F and n are empirical constants depending on both the pollutant (or *adsorbate*) and adsorbent (Table 4.8).

As with all expressions based on chemical equilibrium thermodynamics, the isotherms assume very rapid adsorption. In practice, the rate of adsorption depends on the rates of transfer of adsorbate to the bulk adsorbent surface, and its transfer to the internal surface within the adsorbate pore. These transfer processes are examples of *mass transfer* (Chapter 7), and impact on the overall process *kinetics* (Chapter 5). The kinetics dictate the necessary contact time between water and adsorbent (known as the *empty bed contact time* or *EBCT*), normally in the range of 5–30 minutes for most applications. The EBCT allows calculation of the minimum volume of adsorbent required for a particular duty but this may be increased if the resulting capacity, as determined by Equation 4.50, means that run times are impractical.

EXAMPLE: ADSORPTION

Laboratory adsorption isotherm data for a pesticide to be removed using GAC reveals that $q_e = 5$ mg/g at $c_e = 20\,\mu g/L$, and $q_e = 10$ mg/g at $c_e = 100\,\mu g/L$. Assuming Freundlich behaviour applies, estimate the bed capacity in mg/g GAC, the total bed capacity, the volume treated, and the run time of a 1000 kg column if processing a flow of 200 m^3/hr.

From Equation 4.50: $q_e = K_F c_e^{1/n}$
Taking logs: $\log q_e = \log K_F + (1/n) \log c_e$

This is an example of solution of simultaneous equations (Exercise 2.5). Entering values and converting the concentration units to mg/L for consistency with the Table 4.8 data:

$0.699 = \log K_F + (1/n) \times \log(20/1000) = \log K_F - 1.699(1/n)$

and $\quad 1 = \log K_F + (1/n) \times \log(100/1000) = \log K_F - (1/n)$
Subtracting: $\quad -0.301 = -0.699(1/n)$
So, $\quad 1/n = 0.431$
Substituting into the second of the two simultaneous equations above:

$$1 = \log K_F - 0.431$$

So, $\quad K_F = 27.0$
So, when $c = 10$ µg/L (i.e. 10 mg/m^3 or 0.01 g/m^3), according to Equation 4.50:

$$q_e = 27.0 \times (10/1000)^{0.431} = 3.71 \text{ mg/g} = 3.71 \text{ g/kg}$$

Total bed capacity $=$ Wt $\cdot$ material(kg) $\times q_e$(g/kg) $= 1,000 \times 3.71 = 3710$ g (3.71 kg) pollutant
Total vol. treated $=$ Total bed capacity(g)/concn $\cdot$ (g/m^3) $= 3710/0.01 = 371,000$ m^3
Run time $=$ Volume treated(m^3)/flow(m^3/hr) $= 371,000/200 = 1860$ hours (77 days)

EXERCISE 4.14
How long would the column in the above example last if treating water containing (a) 15 µg/L phenol, or (b) 7.5 µg/L chloroform as the principal contaminant?

4.6.8.2 Ion exchange

In an ion exchange process ions in ions the solution are actually exchanged with ions associated with an insoluble synthetic resin. An example is the well established softening process for removing hardness from water:

$$2(r - \text{Na}) + \text{Ca}^{2+} \rightleftharpoons (r)_2\text{Ca} + 2\text{Na}^+ \tag{4.51}$$

where r represents the resin matrix. The above process allows hardness, normally as calcium and magnesium ions, to be removed from the water and replaced by the less troublesome sodium ion. However, the process can be reversed at much

Table 4.8 Example values of Freundlich constants for a typical GAC.

Compound	K_F, (mg/g)(L/mg)$^{1/n}$	$1/n$
Fluoranthene	664	0.61
Aldrin	651	0.92
DDT	322	0.50
Lindane	285	0.43
1,2-Dichlorobenzene	263	0.38
2,4,6-Trichlorophenol	155	0.40
2,4-Dinitrotoluene	146	0.31
1,4-Dichlorobenzene	121	0.47
Toluene	100	0.45
Nitrobenzene	68	0.43
Tetrachloroethylene	51	0.56
2,4-Dinitrophenol	33	0.61
Trichloroethylene	28	0.62
Phenol	21	0.54
Bromoform	20	0.52
Carbon tetrachloride	11	0.83
Dichlorobromoethane	7.9	0.61
Dibromochloromethane	4.8	0.34
Chloroform	2.6	0.73
Benzene	1.0	1.6

higher sodium concentrations, as a consequence of the Law of Mass Action (Equations 4.23–4.24), allowing the resin to be regenerated using brine solutions (NaCl). In general, exchange in dilute solutions will always take place if the ions in solution are of (a) higher charge and (b) greater weight than those associated with the resin provided, the preference being governed by the *selectivity coefficient*.

The quantity of ion exchange material required between regenerations under a given set of conditions is determined through stoichiometric considerations, and specifically (a) the level of hardness (calcium and magnesium) in the raw water, and (b) the capacity of the ion exchange material for exchanging hardness. These are both normally expressed in equivalents per unit volume (for dissolved hardness) or per unit weight (for the resin capacity), so that the two measurements are comparable on a charge for charge basis. The mass of resin required is then simply given by the total amount of hardness per run divided by the resin capacity:

$$m(\text{kg}) = \frac{\text{run time (h)} \times \text{flow rate (L/h)} \times \text{hardness (eq/L)}}{\text{capacity (eq/kg)}} \tag{4.52}$$

The amount of regenerant can also be calculated along similar principles. The stoichiometric excess of regenerant required to remove the adsorbed hardness for the resin to attain its original capacity is given by the unitless *regeneration ratio*:

$$R = \frac{\text{amount (eq) of regenerant required to regenerate resin}}{\text{amount (eq) of adsorbed hardness on resin}} \tag{4.53}$$

R must always greater than unity, but all ion exchange processes aim to reduce the value of R as much as possible to limit chemical usage and the amount of waste regenerant produced.

Chapter 5
Chemical and biochemical kinetics

5.1 REACTION RATES AND ORDERS

Chemical reactions do not take place instantaneously but advance at varying rates. The way in which the reaction proceeds, and its overall speed, can be used to determine the concentration of a reactant after a specific reaction time. Many reactions take place in a sequence of individual reaction steps all of which proceed at different rates. The slowest reaction is the one which effectively controls the overall rate of reaction and this is thus called the *rate determining step*.

For a general chemical reaction, the *Law of Mass Action* (Equation 4.23) reveals that the rate at which it proceeds is proportional to the concentrations of the reactants. Thus, for the equation:

$$a\mathrm{A} + b\mathrm{B} \Rightarrow c\mathrm{C} + d\mathrm{D},$$

the rate r at which the products are generated and the reactants degrade is given by

$$r = \frac{1}{d}\frac{dc_\mathrm{D}}{dt} = \frac{1}{c}\frac{dc_C}{dt} = -\frac{1}{b}\frac{dc_\mathrm{B}}{dt} = -\frac{1}{a}\frac{dc_\mathrm{A}}{dt} \tag{5.1}$$

where the term dc_X/dt defines the rate of change of the concentration of component X (dc_X) with time (dt). Equations in which such terms appear are generally called *differential equations*.

The rate at which the reaction proceeds is a function of the initial concentrations of the individual components:

$$r = k c_A{}^m c_B{}^n \tag{5.2}$$

where k is the *rate constant* for the reaction, m is the *order* of the reaction with respect to component A and n its order with respect to component B. The rate constant is characteristic of the particular reaction and depends on operating conditions such as temperature, pressure etc. For most reactions m and n will have values of 0, 1 or 2. If $m = 1$ and $n = 0$ the reaction is said to *be zero order with respect to B* and *first order with respect to A*. If both m and $n = 1$ the reaction is *first order with respect to both and A and B* and *second order overall*. Not all reaction kinetics are as simple as this – in some cases m and n are non-integral exponents and in some the exponents are functions of other system parameters.

5.1.1 Rate equations

The relationship between the concentration of the reactants (or products) and time can be obtained by integrating Equation 5.1 with respect to time and with reference to a specific component. Thus, if the rate at which reactant A is consumed is given

© IWA Publishing 2019. Watermaths: Process Fundamentals for the Design and Operation of Water and Wastewater Treatment Technologies
Author: Simon Judd
doi: 10.2166/9781789060393_0051

Table 5.1 Rate equations and half-life expressions.

Reaction order	Rate equation	Rate constant units*	Half-life
0	$c_0 - c_t = kt$	$mg/(L \cdot s)$	$t_{1/2} = c_0/2k$
1	$\ln(c_t/c_0) = -kt$	s^{-1}	$t_{1/2} = \ln 2/k = 0.693/k$
2	$c_t = c_0/(1 + ktc_0)$	$L/(mg \cdot s)$	$t_{1/2} = 1/(kc_0)$

*Assuming concentration is given in mg/L.

generally by the differential equation:

$$r = -\frac{dc_A}{dt} = kc_A^m \tag{5.3}$$

then the kinetic expressions for zero-, first- and second-order reactions can be obtained by substituting values of 0, 1 and 2 respectively for m. Mathematical integration of the resulting differential equations then produces the expressions shown in Table 5.1 below, where the units of k depend on the rate equation. The table also includes the expressions for the reaction *half-life*, which is simply time taken for the concentration of the reactant to decay to half its initial value and is thus a function of the rate constant.

The table shows that:

- If the reaction rate r is independent of the reactant concentration (i.e. zero order), decay is linear with time and the half-life directly proportional to the initial concentration;
- If r is proportional to the reactant concentration (i.e. first order), decay is exponential with time and the half-life is constant;
- If r is proportional to the square of reactant concentration (i.e. second order), the half-life is inversely proportional to the initial concentration.

In fact, many chemical reactions obey first-order kinetics, or else approximate to it (*pseudo first-order*).

EXAMPLE: FIRST-ORDER RATE CONSTANT

What is the rate constant and half-life of a first-order reaction in which the concentration decays by 25% in 5 min?

From Table 5.1, for first-order kinetics (reaction order = 1):

$$\ln(c_t/c_0) = -kt$$

If concentration decays by 25% then $c_t/c_0 = (100-25)/100 = 0.75$
Thus

$$k = -(\ln 0.75)/t = 0.28/5 = 0.0575 \text{ min}^{-1}.$$
$$t_{1/2} = 0.693/k = 0.693/0.0575 = 12 \text{ minutes}.$$

EXERCISE 5.1
If 36% of some micro-organisms die off within 10 min, what is their half-life and how long would it take for 99% of them to die off, assuming first-order kinetics?

EXERCISE 5.2
Nitrogen dioxide (NO_2) decays from a concentration of 7.87 mM at 50 s to 4.81 mM at 200 s. If it is assumed that decay is second order with respect to NO_2, what is the rate constant?

5.2 MICROBIAL KINETICS

5.2.1 Bacterial growth

The growth of single or mixed populations of micro-organisms (i.e. a *culture*) in a submerged system, either freely suspended or immobilised in flocs or films, is very complex. Bacteria reproduce by *binary fission*, i.e. one cell divides into two identical cells. This is called the *cell cycle* or *cell division cycle*. The time taken for this cycle of events is called the *doubling time* – analogous to the half-life in a first-order kinetic reaction. The doubling time is dependent on the species and the conditions of growth, and can range from minutes to days. Since cell growth leads to cell division, producing a daughter cell which is also able to grow and divide, a large population of cells can build up quickly under favourable conditions.

As with reactors (Chapter 8) microbial cultures may be *batch* or *continuous*. In a batch culture system containing an initial, limited amount of *nutrient* (or *medium* which allows microbial growth), once *inoculated* with bacteria (i.e. once the bacteria have been introduced to the medium) the medium composition will change continuously as nutrients are used up and waste products accumulate. This then affects the rate of growth of the micro-organism, the general characteristics of which are shown in Figure 5.1. The culture passes through a number of different phases, all of which take place under *unsteady state* conditions as opposed to *steady state* (Section 6.2.1), and various phases of growth can be identified:

(1) *Lag phase.* Following the inoculation of a culture medium with bacteria, there is a period in which no measurable change in number occurs as the inoculum adapts to the conditions.

(2) *Acceleration phase.* The cells start to reproduce and the rate changes quickly up to the maximum possible without nutrient limitation.

(3) *Unlimited growth phase* (or the *exponential growth phase*). Cells grow at a rate determined by the inter-cellular enzyme reactions, selecting and transporting nutrients as required by the stoichiometry, with the number of cells doubling at a constant rate.

(4) *Deceleration phase.* Different nutrient limitations have differing effects on the physiology of the micro-organisms, and therefore on products that they might synthesise, and this affects the deceleration phase to varying extents.

(5) *Stationary phase.* When the cells can no longer reproduce, due to a limited *substrate* (or food) concentration, high cell concentrations, low partial pressure of oxygen and/or the accumulation of toxic and metabolic end products, the cell population appears static (as with the lag phase), though micro-organisms remain largely viable by existing on stored material.

(6) *Decline phase* or *death phase.* When either the rate of cell death exceeds that of formation of new cells or existing cells use up stored material then a decline in cell population takes place, largely through *autolysis* (whereby cells are destroyed through the action of their own enzymes).

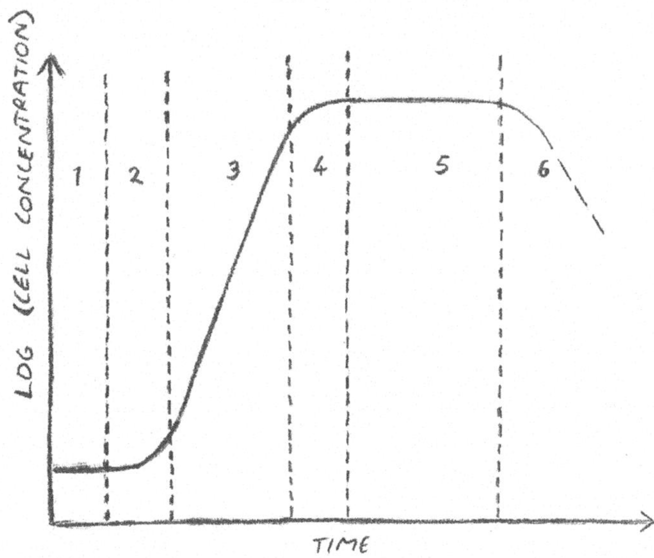

Figure 5.1 Phases of growth for a batch culture.

Table 5.2 Specific growth rate and doubling time values.

Organism	μ_m, h^{-1}	t_d, h
Fungi	0.2	3.5
Yeasts	0.45	1.5
Bacteria	2.0	0.35

5.2.2 Specific growth rate

In a batch reactor (Section 8.3) during the unlimited growth phase (#3 in Figure 5.1) the growth is exponential, and the *specific growth rate* μ can then be defined as the quantity of new biomass formed per unit of original biomass per unit of time t:

$$\ln(x_t/x_0) = \mu t \tag{5.4}$$

or

$$x_t = x_0 \exp(\mu t)$$

where x_t and x_0 respectively represent the number of cells per ml at time t and initially. The above equations are analogous to the first-order chemical reaction (Table 5.1) where exponential decay takes place. As such, the doubling time (t_d) is analogous to the reaction half-life:

$$t_d = \ln 2/\mu. \tag{5.5}$$

Since the growth rate during the exponential phase represents the maximum at which the cells can grow, it is denoted the *maximum specific growth rate* (μ_m, generally in units of h^{-1}) and is a function of the microbial species and the prevailing environmental conditions such as temperature, pH, and substrate and nutrient type and concentration. Typical maximum specific growth rate and corresponding doubling time values for different micro-organism types are given in Table 5.2.

EXAMPLE: SPECIFIC GROWTH RATE

If an exponentially growing culture increases from 1.5×10^5 cells per ml to 5.0×10^8 cells per ml in 6 hours, what is its specific growth rate and doubling time?

From Equation 5.4,

$$\ln(x_t/x_0) = \mu t$$

So,

$$\ln[(5 \times 10^8)/(1.5 \times 10^5)] = \mu \times 6$$
$$\mu = 8.11/6 = 1.35 \text{ h}^{-1}$$

From Equation 5.5, $t_d = \ln 2/\mu = 0.693/1.35 = 0.513$ h.

EXERCISE 5.3
After one day the number of cells per ml for an unlimited exponentially growing culture is 3×10^{18} cells per ml. If the doubling time is 0.4 h what was the initial population?

5.2.3 Nutrient limitation phase

Nutrient limitation occurs when the growth of the organism is restricted by a single nutrient – for example, carbon, nitrogen or phosphorus – resulting in a change in the growth rate (Figure 5.2). The growth curve has a similar form to simple enzyme

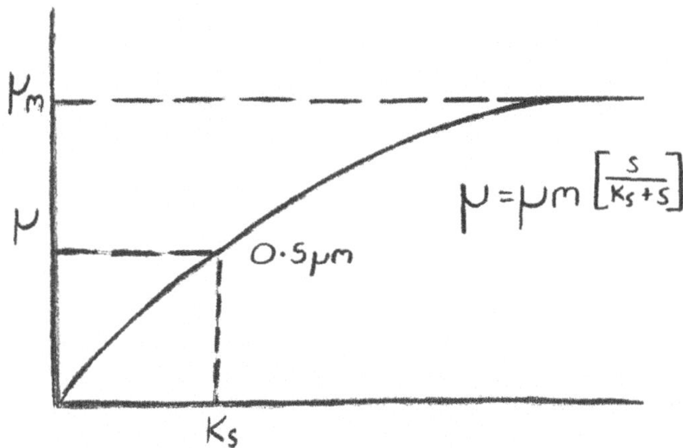

Figure 5.2 Growth rate μ vs. nutrient concentration s, according to Monod kinetics.

reaction kinetics, and can be represented mathematically in a variety of ways but most commonly by the *Monod* equation:

$$\mu = \mu_m \frac{s}{K_s + s} \tag{5.6}$$

where s is the nutrient or substrate concentration and K_s is the half-saturation coefficient for the growth-limiting nutrient.

EXAMPLE: MAXIMUM SPECIFIC GROWTH RATE

What is the maximum specific growth rate if the glucose nutrient concentration is 100 mg C/L (mg of carbon/L), assuming K_s is 5 mg C/L and the growth rate is 0.2 h^{-1}.

In this case $s = 100$ mg C/L, whereas $K_s = 5$ mg C/L.

Since $s \gg K_s$, then according to Equation 5.6: $\mu_m = \mu = 0.2$ h^{-1}.

The value of K_s from Figure 5.2 is that of s when $\mu = \mu_m/2$. Values of K_s for glucose and glycerol range from 2 to 25 mg/L whilst for oxygen it is 0.05 to 0.50 mg/L. K_s provides an indication of the ability of the organism to take up a particular nutrient. At $s \gg K_s$, the value of $s/(K_s + s)$ approaches unity, and in this case $\mu = \mu_m$ and the specific growth rate is at a maximum and independent of the nutrient concentration. At $s \ll K_s$ the Monod equation reduces to:

$$\mu = \mu_m \frac{s}{K_s} \tag{5.7}$$

where μ_m/K_s is a constant, and the specific growth rate is therefore proportional to the nutrient concentration s.

A bacterium growing in a liquid medium providing a mixture of two different nutrients may use one in preference to the other, with the second only being used once the first has been exhausted. During transition between the two sources growth may slow down or even stop as the bacterium adjusts. The resulting growth pattern is characteristic of *diauxic growth*.

5.2.4 Death phase

The kinetics of population decline during the death phase is described by the *Pearl–Verhulst* equation, which assumes the rate to be dependent on the ratio of the population size to the carrying capacity x_f:

$$x_t = \frac{x_f}{1 + e^{c - \mu t}} \tag{5.8}$$

where x_t represents the cell population at time t and c is a constant which relates to the initial concentration x_0 and the carrying capacity:

$$c = \ln\left(\frac{x_f - x_0}{x_0}\right) \tag{5.9}$$

> **EXERCISE 5.4**
> *If a closed culture bacterial population with an initial population of 2.5×10^6 cells/ml, a specific growth rate of 5×10^{-3} min^{-1} and a carrying capacity of 3×10^7, what is the cell population after 2 hours?*

5.2.5 Continuous cultures

Most biological treatment processes used for wastewater treatment operate on a *continuous* basis (Section 6.1), rather than as a *batch* process. There are two principal types of biological process: *attached growth* (or *fixed film*) and *suspended growth*. In the former the micro-organisms, referred to as *biomass*, are fixed onto *media* (plastic or ceramic supporting structures) held in the reactor vessel. For suspended growth processes they are suspended in the tank and recovered either in a separate process by sedimentation (Section 7.5) or else retained in the reactor by a membrane, which allows water out but retains all solid material. Fixed-film processes include the well established *trickling filter* (*TF*, Figure 1.5b). The suspended-growth configuration is the basis of the *activated sludge process* (*ASP*, Figure 1.5a), the *sequencing batch reactor* (SBR), and the more recent *membrane bioreactor* (*MBR*). Some processes combine attached- and suspended-growth configurations, specifically the *moving bed bioreactor* (*MBBR*) or *integrated fixed-film activated sludge* (IFAS) process.

For any biotreatment process the biomass is retained inside it and performs biochemical reactions, such as in Equation 4.22, which result in the degradation of the organic carbon and/or nitrogen compounds in the feedwater passing through the reactor. The water to be treated is fed continuously to the reactor and treated product is withdrawn at the same rate. This continuous culture allows the cells to grow under constant and controlled conditions, such that the models based on batch growth do not apply.

Continuous cultures have two main advantages over batch cultures. Firstly, the cells are kept in a consistent environment, where fluctuations in the concentrations of nutrients and waste products can be minimised. Secondly, by controlling the rate of flow of the medium through the reactor, the rate of bacterial growth can also be controlled. If required, growth can be maintained at the maximum exponential rate. Crucially, because the continuous process contains completely mixed biomass fed with nutrients dissolved in water, one component will be at a limiting concentration compared with the others and thus determine the yield and physiological condition of the cells.

The operation of a biotreatment process based on a continuous culture is therefore dependent on maintaining appropriate conditions for sustaining the biomass whilst retaining it in the bioreactor. For the classical ASP (Figure 1.5a) the biomass particles (or *flocs*) are retained simply by virtue of their size: they are allowed to grow to the point where they are large enough to settle out in a subsequent sedimentation process (Section 7.5). A critical parameter, therefore, is the time the water takes to flow through the tank, or *hydraulic retention time* (*HRT*) θ_w:

$$\theta_w = V/Q \tag{5.10}$$

where V is the reactor volume and Q the feed flow rate. The inverse of θ_w is referred to as the dilution rate. The water must be retained in the reactor for long enough to allow the biomass to grow, and for the organic matter to be degraded. The growth rate of biomass M_b (in kg/h) in the reactor relates to the *yield coefficient Y*, or the proportion of cell material produced by the metabolism of the cell fed with a specified nutrient or substrate:

$$M_b = YQ(s_i - s_R) \tag{5.11}$$

where s_i and s_R respectively represent the concentration in kg $\cdot$ m^{-3} of limiting nutrient in the influent and in the reactor. Since typically $s_R \ll s_i$ the above equation reduces to:

$$M_b \sim YQs_i \tag{5.12}$$

or, substituting for Q from Equation 5.10:

$$M_b/V = Ys_i/\theta_w \tag{5.13}$$

M_b/V being the biomass production rate per unit volume, or *productivity* in kg $\cdot$ m$^{-3} \cdot$ h^{-1}, and Ys_i the reactor biomass concentration (c_B) produced by s_i kg/m^3 of nutrient.

Whilst it is desirable to maintain a rapid biomass growth rate M_b, and thus a high influent nutrient concentration s_i (Equation 5.11), growth can ultimately be constrained by the amount of dissolved oxygen (DO). This applies to any aerobic system, i.e. one where the biomass requires oxygen to survive. For such systems a required specific oxygen uptake rate (q_0) can be determined, and the rate at which oxygen is transferred into the water must then exceed this minimum for growth to be sustained.

The rate of oxygen transfer N in kg/(m^3 h) into the solution is defined by mass transfer (Section 8.2) as:

$$N = k_L a(c^*_{DO} - c_{DO}) \tag{5.14}$$

where $k_L a$ is the absorption coefficient for oxygen in h^{-1} and c_{DO} and c^*_{DO} the mass DO concentrations in the influent and at equilibrium respectively. A maximum cell production rate, $M_{b,\max}/V$ (or $c_{B,\max}/\theta_w$) can then be determined for a given absorption coefficient and permissible minimum DO concentration, $c_{DO,\min}$:

$$\frac{M_{b,\max}}{V} = \frac{k_L a(c^*_{DO} - c_{DO,\min})}{q_0 \theta_w} = \frac{c_{B,\max}}{\theta_w} \tag{5.15}$$

where q_0 is the mass flow rate of oxygen demanded per mass of biomass and has units of kg O$_2$/(kg cells h).

EXAMPLE: HRT FROM OXYGEN MASS TRANSFER

At what HRT must a continuous bioreactor operate to achieve a maximum cell production rate of 2.8 kg cells/(m^3 h), from an oxygen supply of 8×10^{-2} kgO$_2$/(kg cells · h), an equilibrium and minimum DO concentration of 6 and 1 mg/L respectively, and an absorption coefficient of 250 h^{-1}.

From Equation 5.15:

$$M_{b,\max}/V = k_L a(c^*_{DO} - c_{DO,\min})/(q_0 \theta_w)$$

Noting that the units of mg/L for DO concentration equate to kg/m^3 and substituting all values:

$$M_{b,\max}/V = 2.8 \text{ kg/(m}^3\text{h)} = 250 \times (6 - 1) \times 10^{-3}/(8 \times 10^{-2} \times \theta_w) = 15.6/\theta_w \text{ kg/m}^3$$
$$\text{So, } \theta_w = 15.6/2.8 = 5.5 \text{ h}$$

EXERCISE 5.5
What cell mass is generated per day in a 200 m^3 tank from an oxygen supply of 50 gO$_2$/(kg cells · h) at an equilibrium and minimum DO concentration of 5 and 0.5 mg/L respectively, an absorption coefficient of 200 h^{-1} and an 8 h HRT?

Chapter 6
Mass balance

6.1 PROCESSES AND SYSTEMS

A *mass balance* is the determination of mass flow and accumulation in a *system*. The system is some volume of space enclosed by an arbitrary boundary inside of which a *process* takes place causing a change in state or physical and/or chemical composition of matter. The process may be a simple blending of two streams of water containing two different concentrations of some dissolved or suspended material, or a vessel in which chemicals react to form products. A mass (or material) balance is thus a simple sum based on the fundamental principle of matter being conserved. This then entails the determination of mass flows, usually in kg/hr, into and out of the system boundary.

The system for which the mass balance is performed may be either *closed*, if material and energy cannot be exchanged with the surroundings, or *open* if it can. The types of process within a system determine the nature of the change in mass with time. There are three common types of processes in water and wastewater treatment: *batch*, *fed batch*, and *continuous*. In a batch process the materials are added to the reactor (Chapter 8) at the start and the product(s) only removed when the reaction has reached some end point, whereas in a continuous process materials flow to and from the reactor during the reaction. For a fed batch process materials are added to material already in the reactor, and the product(s) removed at end as with the batch process.

Batch processes tend to be used when volumes requiring treatment are small, and/or the product has high added value. Some sludge treatment processes in the water industry are batch (filter presses and drying beds, for example). Other sludge processes are semi-continuous, with material added and product removed at intervals. In water and wastewater treatment most processes are designed to be continuous and to operate at *steady state*, i.e. a condition where none of the bulk properties (concentration, temperature, pressure, etc) vary with time. Such processes range from simple physical sedimentation (Section 7.5) to classical biotreatment processes such as the activated sludge process (Section 6.2.4) and trickling filters and advanced processes such as advanced oxidation and reverse osmosis. Steady state is to be distinguished from equilibrium (Section 4.6), which implies no overall change with time – i.e. no change in the reactants and products. For a continuous process operating at steady state, the rate at which reactants are being fed into the process is balanced by the rate at which generated products are removed from it.

6.2 MASS BALANCE CALCULATIONS
6.2.1 Steady state

Since the mass of a system is always conserved, the total mass of material within a system is given by:

$$\text{mass flow in} = \text{mass flow out} + \text{mass rate of accumulation.} \tag{6.1}$$

If there is no accumulation within the system, i.e. it is at steady state, then the flow of mass in is equal to the flow out. If an example of a simple sedimentation tank is taken (Figure 6.1a) then a mass balance can be conducted around the tank which forms the system, and the boundary (dotted line in figure) drawn accordingly. If the sludge is then dewatered in a subsequent process (Figure 6.1b) then, if desired, the system boundary (System 3 in Figure 6.1b) could be extended to include that process and the number of streams expanded by one to include the sludge supernatant or filtrate. The sludge processing could otherwise be considered separately (System 2). If the filtrate were to be returned to the inlet of the sedimentation tank (i.e. just upstream of System 1, but within System 3) then the number streams entering and leaving the System 3 would then be reduced back down to three. It would therefore appear similar to the original sedimentation tank (System 1), but defined by different flows.

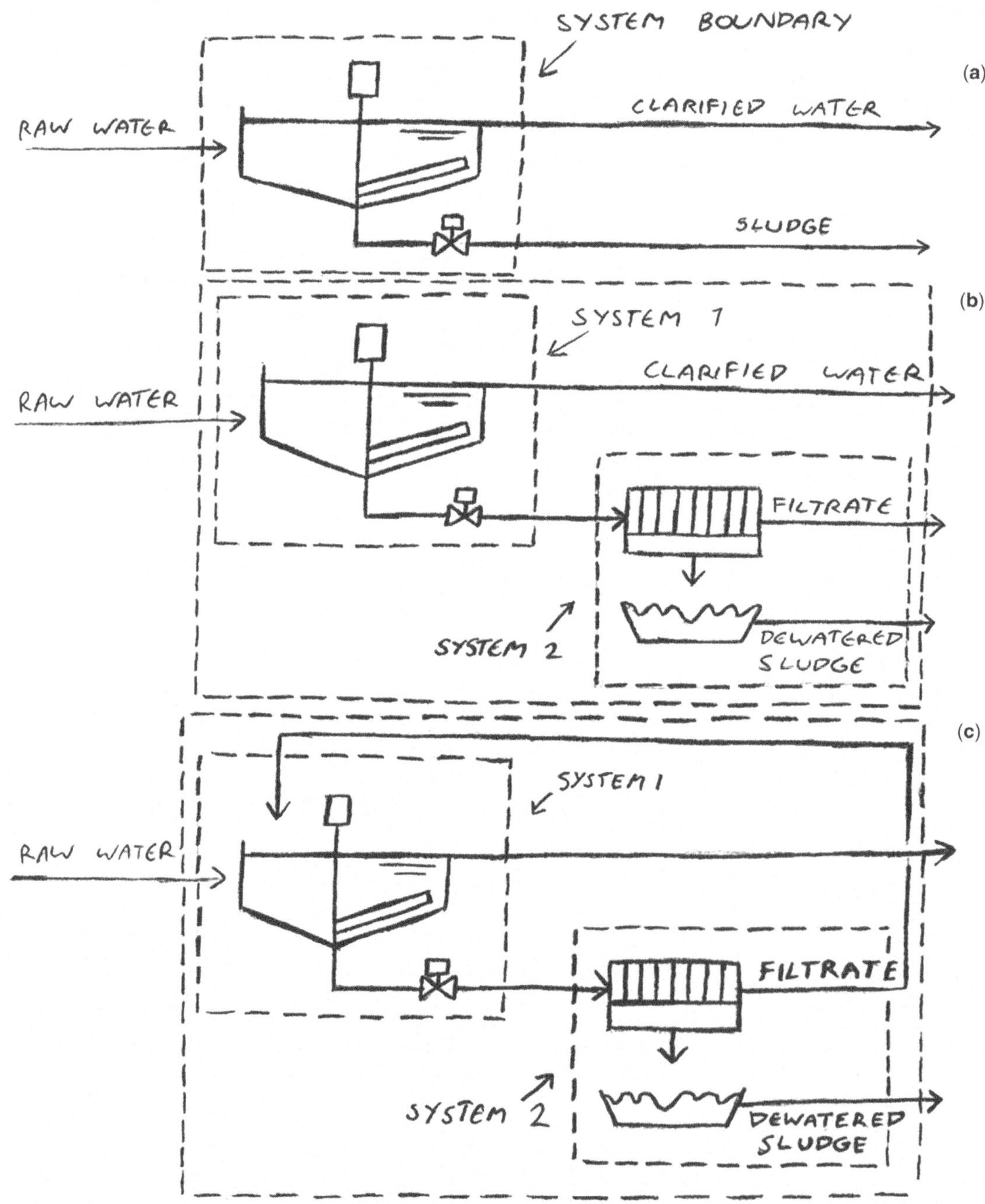

Figure 6.1 (a) Sedimentation tank, (b) sedimentation tank with downstream sludge dewatering, and (c) with recycling of the filtrate.

In all cases, the detail of what happens inside the boundary is unimportant, other than how this impacts on the mass flow of solids in the streams (such as through chemical reaction, Section 6.2.3). The principle can be extended to flow of moles of suspended and dissolved materials, even when a chemical transformation and/or a change of phase (liquid to solid, liquid to gas, etc) arises in the reactor, since matter is always conserved.

If flows are intermittent, as is often the case for operating water and wastewater treatment plants, they need to be averaged. For example for the physical separation processes of media and membrane filtration, the waste solids stream is generated by periodic washing of the filter by applying washwater in the opposite direction to the filtration flow – an action known as *backwashing* or *backflushing*. Since the process has to be stopped for backwashing to take place, and backwashing also uses some of the product water, the overall production rate is less than the filtration rate.

EXAMPLE: INTERMITTENT DISCHARGE FROM A TANK

If sludge is intermittently discharged from a tank at a flow rate of $Q_{intermit.}$ of 500 m^3/h for a period t of 2 minutes each 20 minutes (t_{cycle}), what is the net discharge per hour?

Discharge rate = cycles/hr $\times$ $Q_{intermit.}$ $\times$ fraction of hour over which $Q_{intermit.}$ applies

$$Q_{discharge} = (60/t_{cycle}) \times Q_{intermit.} \times (t/60) = Q_{intermit.}t/t_{cycle} = 500 \times 2/20 = 50 \ m^3/h.$$

EXERCISE 6.1
Water is pumped to a tank by a pump which operates twice each hour for 15 minutes at a flow rate of 25 L/s. What is the average hourly flow into the tank in m^3/h?

EXERCISE 6.2
Water flows through a membrane filter at a rate of 12 L/s. Every 18 minutes it is backflushed for 90 seconds at a rate of 16 L/s and then takes another 30 s to go back into service. What is the net production rate as a proportion of the filtration flow rate?

In all cases, flows should be converted to consistent units prior to conducting the mass balance. Water flows in municipal water works are often expressed in L/s. It is convenient to convert all units to mass flows of the dissolved/suspended material and of the stream itself, normally in kg/h or te/h.

For the mass balance the mass flow rate M of the material in, for example, kg/h is simply mass concentration C in kg/m^3 or kg/te if the density is assumed to be 1 te/m^3 multiplied by the flow rate of the stream Q in either m^3/h or te/h. It thus follows that:

$$\Sigma(Q_{in}C_{in}) = \Sigma(Q_{out}C_{out}) + \Sigma M_{accumulation} \tag{6.2}$$

Since the water is also conserved, the mass or volumetric flows must also balance:

$$\Sigma Q_{in} = \Sigma Q_{out} \tag{6.3}$$

In the case of the sedimentation tank (or clarifier) in Figure 6.1a:

$$Q_1 C_1 = Q_2 C_2 + Q_3 C_3 \tag{6.4}$$

and $\quad Q_1 = Q_2 + Q_3$ $\tag{6.5}$

where the subscripts 1, 2 and 3 refer to the feed, clarified product and sludge solids concentrations respectively. Often the product stream Q_2 needs to be calculated. Putting $Q_3 = Q_1 - Q_2$ (Equation 6.5), and substituting this term into Equation 6.5 and rearranging:

$$Q_2 = Q_1(C_3 - C_1)/(C_3 - C_2). \tag{6.6}$$

EXAMPLE: MASS BALANCE ACROSS A CLARIFIER

Water containing 300 mg/L of suspended solids and flowing at 2 MLD (megalitres/day) is to be treated by sedimentation (Figure 6.1a). The process produces treated water containing 20 mg/L of suspended solids and a sludge product with a solids concentration of 0.8% by weight. How much sludge and clarified water is produced?

Concentrations in water treatment are traditionally measured in mg/L, where $1\,\text{mg/L} = 10^{-3}\,\text{kg/m}^3$. Since water has a density of $1000\,\text{kg/m}^3$, converting flows from m^3/h to kg (or te) per h and concentrations from mg/L into mg/kg or (g/te) is straightforward: 20 mg/L equates to 0.02 kg/te. Similarly, 0.8% by weight equates to 0.8 g per 100 g, which is the same as 8 g per 1000 g and so 8 kg/te.

So, converting all quantities to kg, te and h and substituting into Equation 6.6:

$$Q_2 = (2000/24) \times (8 - 0.3)/(8 - 0.02) = 83.33 \times 7.7/7.98 = 80.41\,\text{te/h},\ 1.930\,\text{MLD}$$

And so $\quad Q_3 = Q_1 - Q_2 = 83.33 - 80.41 = 2.92\,\text{te/h}.$

EXAMPLE: MASS BALANCE ACROSS A FILTER PRESS

The sludge from the clarifier from the above process is to be dewatered to 45% dry solids by filter pressing (Figure 6.1b), producing a filtrate of 500 mg/L solids. Calculate the flow of filtrate and dewatered solids, and so the overall efficiencies for water and solids recovery.

In this case the mass balance parameter values are:

$$\text{Feed}:\quad Q_1 = 2.92\,\text{m}^3/\text{h} = 2.92\,\text{te/h}$$
$$C_1 = 0.8\,\text{g}/100\,\text{g} = 8\,\text{g/kg} = 8\,\text{kg/te}$$

$$\text{Filtrate:}\ C_2 = 500\,\text{mg/L} = 0.5\,\text{kg/te}$$
$$\text{Sludge:}\ C_3 = 45\,\text{g}/100\,\text{g} = 450\,\text{kg/te}$$

Equation 6.6 can again be used to determine the filtrate flow:

$$Q_2 = 2.92 \times (450 - 8)/(450 - 0.5) = 2.87\,\text{te/h}$$

So, $\quad Q_3 = Q_1 - Q_2 = 2.920 - 2.871 = 0.049\,\text{te/h}.$

For the overall mass balance based on the wider boundary (System 3) shown in Figure 6.1c, encompassing the feed, filtrate and dewatered sludge:

$$\text{Mass flow in} = 83.33 \times 0.3 = 25.0\,\text{kg/h}$$

$$\text{Mass flow, sludge product} = 0.049 \times 450 = 21.9\,\text{kg/h}$$

Which means that the process recovers 21.9/25.0 or 88% of the solids from the feedwater in the dewatered sludge.

If the two clarified product streams were to be blended then the fraction of water recovered would be:

$$\%\text{recovery} = (Q_{2,\text{stage 1}} + Q_{2,\text{stage 2}})/Q_1 = (80.41 + 2.87)/83.33 = 99.90\%$$

The concentration in this stream would be the mass flow $(M_{\text{feed}} - M_{\text{dewatered sludge}})$ divided by the volume flow $(Q_{2,\text{stage 1}} + Q_{2,\text{stage 2}}$, assuming a density of $1000\,\text{kg/m}^3)$. Thus:

$$C_{\text{product, overall}} = (25 - 21.9)/(80.41 + 2.89) = 0.037\,\text{kg/m}^3,\ \text{or } 37\,\text{mg/L}$$

The alternative to product blending would be to return the clarified stream from the filter press to the inlet of the sedimentation tank (Figure 6.1c). This is an example of recycling, and creates a rather more complicated mass balance problem (Section 6.2.2).

EXERCISE 6.3

50 m³/day of sludge containing 10 kg/te of solids is to be dewatered to 40% dry solids by filter pressing. Calculate the daily mass (in kg) of dry solids in the dewatered sludge and the volume of filtrate generated assuming the filtrate contains 50 mg/L of suspended solids.

EXERCISE 6.4

A wet sludge originally containing 71% of water by weight is dried to remove 60% of the water. What weight of water is removed per kg of wet sludge, and what is the solids content (% by weight) of the dried sludge?

Mass balances apply to all components, both suspended and dissolved. Treatment of water can involve a change of phase, either through chemical reaction (Section 6.2.3) or physical means. Levels of dissolved materials can be decreased by adsorption (onto activated carbon or ion exchange materials, Section 4.6.8), membrane filtration (using *dense membrane* processes such as reverse osmosis or nanofiltration), chemical reaction or simply by blending. In all cases, for steady-state operation the mass balance principles as depicted in Equations 6.2 and 6.3 can be applied.

EXAMPLE: SEAWATER EVAPORATION

Sea water of density 1.01 g/ml containing 28 g/L of sodium chloride is evaporated to produce salt. If the salt production required is 2 tonnes per day and the plant operates for 8 hours a day, what seawater volume flow has to be processed during the operating cycle?

Sea water density $= 1.01 \, g/cm^3 = 1010 \, kg/m^3$

So, mass concentration (remembering $1 \, g/L = 1 \, kg/m^3$) $= 28/1010 = 0.0277 \, kg/kg$

Mass flow salt M (kg/h) $=$ mass concentration (kg/kg) $\times$ mass flow seawater Q (kg/h) $= 0.0277 \, Q$

This equates to the salt production, $M = 2 \, te$ every $8 \, h = 2000 \, kg/8 \, h = 250 \, kg/h$

So, $0.0277 \, Q = 250$, and thus $Q = 9025 \, kg/h$, or $9.03 \times 1000/1010 = 8.94 \, m^3/h$

EXERCISE 6.5

A solution of potassium chloride of density 1.12 g/mL containing 45 g/L of the salt is evaporated to produce a dry product. If the potassium chloride production required is 5 tonnes per day and the plant operates for 12 hours each day, what is the solution flow (mass and volume) which has to be processed during the operating cycle?

EXAMPLE: BLENDING

---

100 m³/h of surface water containing 110 mg/L of sodium is mixed with borehole water containing 10 mg/L of sodium. If the mixed stream is to have a sodium concentration of no more than 80 mg/L what is the minimum flow of borehole water required?

In this case there are two feed streams and a single product stream, and so the mass and flow balances equations are:

$$Q_1 C_1 + Q_2 C_2 = Q_3 C_3$$

and $$Q_1 + Q_2 = Q_3$$

where $C_1 = 110$ g/m^3, $Q_1 = 100$ m^3/h, $C_2 = 10$ g/m^3, $C_3 = 80$ mg/L or less, and Q_2 is the minimum flow required for the borehole water.

So, $Q_3 = 100 + Q_2$. Substituting this into the mass balance equation:

$$(100 \times 110) + (10 \times Q_2) = 80 \times (100 + Q_2)$$

Rearranging:

$$Q_2 = [(100 \times 110) - (80 \times 100)]/(80 - 10) = 42.9 \, \text{m}^3/\text{h}$$

EXERCISE 6.6

10 m^3/h of water containing 80 mg/L of nitrate is mixed with surface water containing 10 mg/L of nitrate. If the mixed stream is to have a nitrate concentration of no more than 50 mg/L what is the minimum flow of surface water required?

Dense membrane processes (reverse osmosis, RO, and nanofiltration, NF) are usually configured as *stages*, where the second stage receives the waste concentrate (or *retentate*) stream from the first stage to recover more of the water – analogous to the filter press receiving the sludge stream from the sedimentation tank in Figure 6.1b. They are based on *modules*, where each module contains a number of membrane *elements* connected in series. Analogous to a solids separation process, there is normally a single feed stream, a purified product stream (the *permeate* in the case of a membrane process) and a *concentrate* waste stream (Figure 1.8(o)). This applies to the element, the module and the overall process depending on where the boundary for the mass balance is drawn.

Within the module, the concentrate stream from one element is passed on to the feed of the next, in the same way to the concentrate from one stage being passed on to the feed of the next (Figure. 6.2). The general expression for the recovery R from a series of n elements in a module, or through n stages in a membrane process, is:

$$R_{\text{overall}} = Q_P/Q_F = 1 - (1 - \alpha)^n \tag{6.7}$$

where α is the recovery per element (in a module) or per stage (in a process). There are normally 4–6 elements in an RO/NF membrane module and 2–3 stages in an RO/NF process (normally called an *array*).

Equation 6.7 applies to any stage-wise process where there is separation of some kind at each stage. It is usually simpler to conduct calculations based on the retentate stream:

$$1 - R_{\text{overall}} = Q_R/Q_F = (1 - \alpha)^n \tag{6.8}$$

Since the membranes do not remove all of the dissolved matter, the mass balance across the module can be conducted in the same way for the dissolved solids and for suspended solids in the case of a solids separation process such as sedimentation.

EXAMPLE: DENSE MEMBRANE PROCESS

Water containing 300 mg/L of salt flows into a two-stage nanofiltration process at a rate of 500 m^3/h, where the concentrate from the first stage is fed to the second stage (Figure 6.2). At each stage:

- *The recovery (the ratio of the flow of product to that of the feed) is 0.45, and*
- *The salt passage (the ratio of the product concentration to the feed concentration) is 0.15 per stage.*

The product (or permeate) from each stage is combined to provide the overall flow.

What is the overall recovery and salt passage?

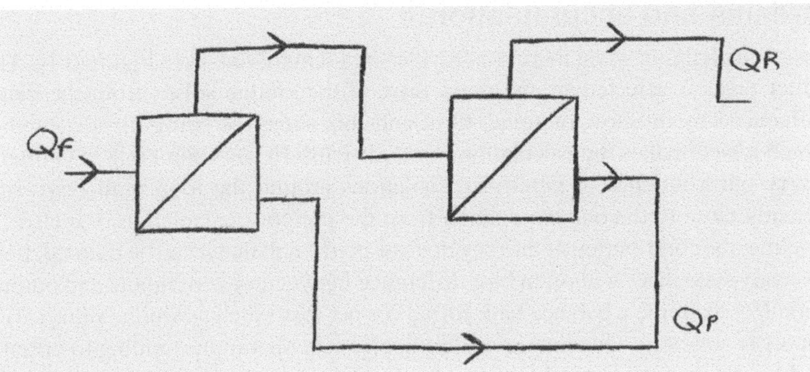

Figure 6.2 A two-stage membrane process.

Recovery $R = Q_{permeate}/Q_F = 0.45$, where $Q_F = 500\,\text{m}^3/\text{h}$ for the first stage.

For each stage, $Q_{feed} = Q_{permeate} + Q_{concentrate}$,

So, $Q_{concentrate, Stage 1} = Q_F \times (1 - R) = 500 \times (1 - 0.45) = 275\,\text{m}^3/\text{h}$

The flow of concentrate flows to the second stage where is further concentration by a factor of $(1 - R)$, thus:

$$Q_{concentrate, Stage 2} = Q_R = Q_F \times (1 - R)^2 = 500 \times (1 - 0.45)^2 = 151\,\text{m}^3/\text{h}$$

The product (or *permeate*) from each stage is combined to provide the overall permeate flow Q_P, where:

$$Q_P = Q_F - Q_R = 500 - 151 = 349\,\text{m}^3/\text{h}$$

So, the overall recovery is $349/500 = 0.698$ (or 69.8%). This is the same as value as that given as a function of the recovery per stage: $1 - (1 - 0.45)^2$.

The salt passage per stage $= C_{permeate}/C_{feed} = 0.15$

So, in Stage 1, $C_{permeate} = 0.15 \times 300 = 45\,\text{mg/L} = 0.045\,\text{kg/m}^3$, cf. $C_F = 0.3\,\text{kg/m}^3$, and so

since $Q_F C_F = Q_{concentrate, Stage 1} C_{concentrate, Stage 1} + Q_{permeate, Stage 1} C_{permeate, Stage 1}$

$\qquad 500 \times 0.3 = 275 \times C_{concentrate, Stage 1} + (500 - 275) \times 0.045$

then, $C_{concentrate, Stage 1} = (150 - 10.1)/275 = 0.509\,\text{kg/m}^3$,

and total mass flow of salt in permeate for the second stage is:

$$M_2 = Q_{permeate, Stage 2} C_{permeate, Stage 2} = (0.45 \times 275) \times (0.15 \times 0.509) = 9.44\,\text{kg/h}$$

Adding this to the permeate salt mass flow in Stage 1:

$$M_P = M_1 + M_2 = (500 - 275) \times 0.045 + 9.44 = 19.57\,\text{kg/h}$$

So, the blended permeate concentration C_P in mg/L, is

$$M_P/Q_P = 1000 \times (19.57/349) = 56\,\text{mg/L}$$

which means that the overall salt passage is $56/300 = 0.186$ – rather higher than the value of 0.15 for each stage because the second stage receives a higher concentration of salt than the first and so quantitatively allows more salt through.

6.2.2 Recycled streams and accumulation

Mass balances become more challenging when there is a recycle stream involved, as in Figure 6.1c. The mass balance conducted previously around the filter press is affected by the mass flow of the sludge solids from the clarifier. The balance around the clarifier is in turn influenced by the flow of solids to it, and this stream is being blended with the filtrate from the filter press. A loop is thus created which makes the calculation more difficult. In such situations it is often necessary to estimate the recycle flow and then carry out a number of *iterative* calculations around the loop until *convergence* occurs, i.e. until the calculated value is sufficiently close to the one determined from the previous calculation (see also Section 3.4). In the case of the filter press filtrate recycling, the contribution of the recycle flow is so small that it can be ignored, but this is not always the case.

If the system is not at steady state there will often be a difference between system inputs and outputs, which means that there is either a net accumulation (for example a balance tank filling) or net loss (such as sludge removal) of material. In such a case the final term in Equation 6.1 is non-zero. This may arise during physical operations (adding to or removing from the system) or from chemical reactions. In both cases non-steady state operation implies a batch, rather than continuous, process.

> **EXERCISE 6.8**
> *Water is pumped from the tank in Exercise 6.1 at a constant rate of 43.5 m^3/h. What is the rate of accumulation of water? If the tank has a working volume of 12 m^3, how long will it be before the tank overflows?*

6.2.3 Chemical reaction

The generalised mass balance statements apply when there is no chemical reaction occurring within the system, i.e. no transformation of mass. Should chemical reaction be involved a particular component may be destroyed or created within the system, and for a batch process this will respectively lead to depletion or accumulation of that component. However, if the reacting components are replaced and the chemically generated products removed on a continuous basis, then there may be no accumulation within the system. In such a case the mass balance becomes:

$$\text{mass flow in} + \text{mass generated} = \text{mass flow out} + \text{mass consumed} \tag{6.9}$$

In the case of chemically generated or depleted components, it is important to understand the nature of the chemical reaction and, specifically, its stoichiometry (Section 4.1). A simple example is precipitation, where a dissolved contaminant such as the ferrous ion (Fe^{2+}) is removed by chemically converting it to a substantially insoluble salt (Section 4.5.2.2). Water containing dissolved iron can be aerated to oxidise the ferrous ion to ferric hydroxide which is precipitated and so removed by clarification, the chemical reaction being:

$$2Fe^{2+} + \tfrac{1}{2}O_2 + 5H_2O = 2Fe(OH)_3\downarrow + 4H^+ \tag{6.10}$$

The system thus resembles the schematic below (Figure 6.3). If the feed concentration of Fe^{2+} is 10 mg/L, then at the Fe molar mass of 56 (Table 4.1) this equates to a molar concentration of $10/56 = 0.179$ M. The same molar concentration of $Fe(OH)_3$ precipitate is generated, its molecular weight being $56 + 3 \times (16 + 1) = 107$. The product $Fe(OH)_3$ concentration in mg/L, assuming the reaction goes to completion, is therefore $107 \times 0.179 = 19.1$ mg/L. This represents the feed suspended solids concentration to the clarifier, around which the mass balance method can be applied in exactly the same as

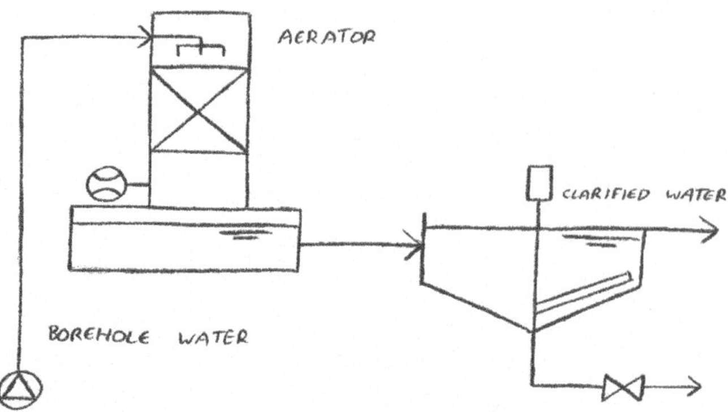

Figure 6.3 Schematic for chemical oxidation of Fe^{2+} followed by sedimentation.

before if steady state is again assumed. A mass balance for iron across the aerator shows that the concentration of iron does not change between the feed and the product streams, but that dissolved Fe^{2+} levels are reduced and $Fe(OH)_3$ solids commensurately generated.

EXAMPLE: FERROUS OXIDATION AND REMOVAL

A 360 m^3/d flow of water containing 10 mg/L of Fe^{2+} is treated by an aeration tower, which oxidises 90% of the Fe^{2+} to insoluble $Fe(OH)_3$ which is then removed by sedimentation to produce a 20 kg/h sludge product flow of 1.2 wt% (Figure 6.3). What is the residual iron concentration?

According to Equation 6.4:

$$Q_1 C_1 = Q_2 C_2 + Q_3 C_3$$

In this case 90% of the Fe^{2+} is converted to $Fe(OH)_3$ with a stoichiometry of $1:1$. The corresponding mass ratio, according to Table 4.2 is $56:(56 + 3 \times 17) = 56:107$ or $1.92:1$. The flow and concentrations for the feed and waste streams are thus:

$$Q_1 = 360\,m^3/d = 15\,te/h$$
$$C_1 = 0.9 \times 10 \times 1.92/10^3\,g/L = 0.0173\,kg/te \text{ (since 1 mM } Fe^{2+} \equiv 1.92\,mM\,Fe(OH)_3)$$
$$Q_3 = 20\,kg/h = 0.02\,te/h$$
$$C_3 = 1.2\,wt\% = 12\,kg/te$$

and C_2 is the ferric hydroxide concentration in the clarified product.

Since from Equation 6.5 $Q_2 = Q_1 - Q_3$, substituting this into Equation 6.6:

$$Q_1 C_1 = (Q_1 - Q_3)C_2 + Q_3 C_3$$

So, $\quad C_2 = (Q_1 C_1 - Q_3 C_3)/(Q_1 - Q_3)$ as before

$$= ((15 \times 0.0173) - (0.02 \times 12))/(15 - 0.02) = 1.30 \times 10^{-3}\,kg/te \equiv 1.28\,mg/L\,Fe(OH)_3$$

Converting to the concentration as Fe:

$$C_2 = 1.30/1.91 = 0.68\,mg/L \text{ as Fe.}$$

This concentration has to be added to the residual ferrous ion (Fe^{2+}) concentration from the oxidation tower. If this all remains dissolved, and so ends up in the clarified product stream, then:

$$C_{2,\text{residual FeII}} = (1 - 0.9) \times 10 = 1.00\,mg/L$$

So, $\quad C_{2,\text{tot Fe resid.}} = 1.00 + 0.68 = 1.68\,mg/L$

EXAMPLE: ACID NEUTRALISATION

Hydrochloric acid reacts with lime as follows:

$$2HCl + Ca(OH)_2 \Rightarrow CaCl_2 + 2H_2O$$

How much lime in kg/h is required to completely neutralise 100 m^3/h of hydrochloric acid containing 7 wt% HCl if the acid solution has a density of 1025 kg/m^3, and what concentration in g/L of calcium chloride is produced?

The molecular weights of lime and hydrochloric acid, according to Table 4.2, are given by:

$$Ca(OH)_2 = 40 + 2 \times (16 + 1) = 74\,g$$
$$HCl = 1 + 35.5 = 36.5$$

According to the chemical equation for the reaction, one mole of $Ca(OH)_2$ is required to neutralise two moles of HCl, and so the mass ratio is $74/(2 \times 36.5) = 1.014$.

The concentration of acid is given as 15% by weight and its density (ρ) as 1025 kg/m^3. The mass flow of acid in kg/h is given by:

$$M = Q(m^3/h) \times C(\text{kg HCl/kg water}) \times \rho(\text{kg water}/m^3\text{water})$$
$$= 100 \times (7/100) \times 1025 = 7175$$

So, required mass flow of Ca(OH)$_2$ = 1.014 $\times$ 7175 = 7275 kg/h.

The calcium chloride concentration in kg/m^3 (and so g/L) is given by the ratio of the mass flow of CaCl$_2$ in kg/h to the flow of water in m^3/h. In this case, water is generated by the reaction: according to the equation one mole of CaCl$_2$ (molecular weight = 40 + (2 $\times$ 35.5) = 111) and two moles of water (molecular weight = (2 $\times$ 1) + 16 = 18) are generated for every mole of lime dosed. So:

Mass flow CaCl$_2$ generated = (111/74) $\times$ 7275 = 10913 kg/h
Mass flow water generated = (2 $\times$ 18/74) $\times$ 7275 = 3539 kg/h

The mass flow of feed water is given by the total mass flow minus the acid mass flow:

Mass flow of feed water = (1025 kg/m^3 $\times$ 100 m^3/h) − 10913 = 91587 kg/h

So, the total mass flow of water generated is:

Feed + generated flow = 91587 + 3539 = 95126 kg/h, or 95.12 m^3/h for a density of 1000 kg/m^3

Therefore the product calcium chloride concentration is given by:

Mass flow CaCl$_2$/volume flow water = 10913/95.12 = 115 kg/m^3(org/L).

EXERCISE 6.9
Sulphuric acid reacts with lime to produce calcium sulphate. How much lime in te/h is required to neutralise 350 m^3/h of dilute sulphuric acid containing 15% H$_2$SO$_4$ if the acid solution has a density of 1100 kg/m^3? What concentration of dissolved calcium sulphate is generated, assuming it is completely soluble? How much would be precipitated if the solubility limit was 2 g/L?

EXERCISE 6.10
A laundry uses 300 m^3/day of mains water for washing industrial workwear. Detergents and dirt from the wash loads add 600 kg/day of COD to the water. If 10% of the water is lost by evaporation, calculate the COD of the wastewater discharge in mg/L.

The COD concentration in the wastewater discharge is reduced to 1500 mg/L by ultrafiltration (UF), a membrane process. The UF produces a permeate (80% of the flow) and retentate (20%), the permeate COD concentration being 70% of the UF influent. Calculate the permeate and retentate flow rates and COD concentrations.

The retentate waste stream is then treated by adding 270 mg/L of ferric chloride (FeCl$_3$) to precipitate ferric hydroxide and remove 85% of the COD as suspended solids (SS), where the COD:suspended solids weight ratio is 2:1. The coagulated stream is clarified in a settling tank which produces sludge at 2% solids and a clarified water stream of 20 mg/L TSS. Calculate the clarified water and sludge flows in te/day.

6.2.4 Biological processes

Biological processes are the single most important type in treating wastewater. In this process, micro-organisms are used to degrade organic carbon, through reactions such as that given in Equation 4.22. In the activated sludge process (ASP) the micro-organisms form clusters, known as *flocs*, and make up the major part of the solids in the tank. The solids are collectively referred to as the *mixed liquor suspended solids* (*MLSS*). The micro-organism (or *biomass*) content of the MLSS is normally quantified as the volatile fraction, and is hence referred to as the *mixed liquor volatile suspended solids* or *MLVSS*.

The tank is fed with the wastewater, containing the organic material, and air, which provides oxygen to the micro-organisms. The established form of the ASP employs a conventional clarifier (Figure 6.1a) to separate the solids from the treated water. This configuration of the process, where the biological solids are recovered by clarification and fed back to the biological process

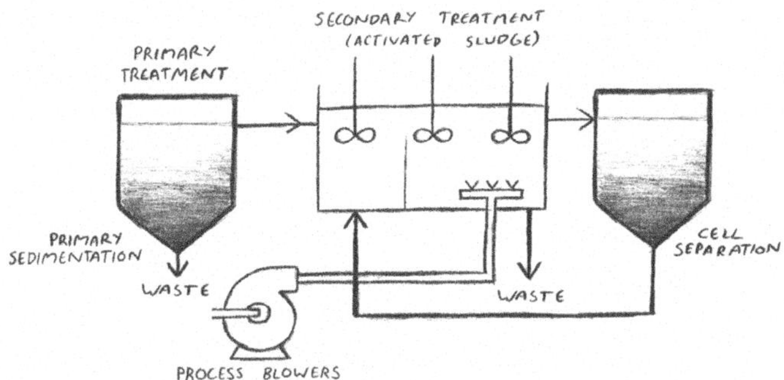

Figure 6.4 The three-stage CAS-based wastewater treatment process.

tank (Figure 6.4), is the most widely implemented. It is often referred to as the *conventional* (or *classical*) *activated sludge process* (*CAS*).

In the case of biological treatment the mass flow of solids through the process tank is defined by the *solids retention time* (*SRT*, usually denoted θ_s). This refers to the time the solids spend in the tank, as distinguished from the time the water spends in the tank which is the *hydraulic retention time* (*HRT*, θ_w). If there are no residual solids in the treated water, i.e. $C_2 = 0$, and the solids concentration in the sludge waste stream (C_3) is the same as the concentration in the tank, then θ_s is simply the ratio of the tank volume V_{tank} to the outlet solids flow rate (Q_3):

$$\theta_s = V_{tank}/Q_3 \tag{6.11}$$

This is the case when the retention of the solids in the tank is via a membrane, such as in a membrane bioreactor (MBR). For the CAS process, where sedimentation is used to retain the solids, there is always a residual solids concentration in the treated water. Applying a mass balance then yields:

$$\theta_s = \frac{V_{tank}X}{Q_2C_2 + Q_3C_3} \tag{6.12}$$

where X is the MLSS concentration, Q_2 and C_2 refer to the flow rate and solids concentration in the treated effluent, and Q_3 and C_3 the corresponding parameters for the waste sludge stream.

EXAMPLE: SRT DETERMINATION

Water flows through a 15,000 m³ activated sludge tank containing mixed liquor suspended solids at a concentration of 3.5 g/L and providing a hydraulic residence time of 7.5 h. If 2.5% of the feed flow forms the sludge waste, the waste solids concentration is 15% more than that of the MLSS, and the suspended solids concentration in the treated effluent is 25 mg/L. What is the SRT?

The SRT is given by Equation 6.12:

$$\theta_s = \frac{V_{tank}X}{Q_2C_2 + Q_3C_3}$$

where Q_3 is 2.5% of Q_1, Q_1 being the feed flow, and Q_2 is therefore 97.5% of Q_1. Q_1 is given by the volume per unit time:

$$Q_1 = V_{tank}/\text{HRT} = 15{,}000/7.5 = 2{,}000 \, \text{m}^3/\text{h}.$$

And thus, $Q_3 = 0.025 \times 2{,}000 = 50 \, \text{m}^3/\text{h}$, and $Q_2 = 0.975 \times 2{,}000 = 1950 \, \text{m}^3/\text{h}$.

The sludge waste concentration C_3 is 15% more than the MLSS concentration:

$$C_3 = 1.15 \times 3.5 = 4.03 \, \text{g/L} = 4.03 \, \text{kg/m}^3$$

The treated water concentration C_2 is 25 mg/L $= 0.025 \, \text{kg/m}^3$

$$\text{So} \quad \theta_s = \frac{15{,}000 \, \text{m}^3 \ \times \ 3.5 \, \text{kg/m}^3}{(1950 \, \text{m}^3/\text{h} \ \times 0.025 \, \text{kg/m}^3) + (50 \, \text{m}^3/\text{h} \ \times 4.03 \, \text{kg/m}^3)} = \frac{52{,}500}{48.8 + 201.5} = 210 \, \text{h} = 8.75 \, \text{d}$$

EXERCISE 6.11
What would be the new treated suspended solids concentration if the SRT was extended to 9.5 days and the sludge waste concentration increased to 4.2 kg/m^3, assuming all other parameters to remain unchanged?

A further critically important parameter in biological treatment is the *food-to-micro-organism ratio* (*F:M* or *F/M ratio*). This parameter defines the ratio of the substrate (or food) entering the process tank to the mass of microbiological material (or biomass) in the tank. Because biological processes operate through the use of micro-organisms which biodegrade the organic material in the feedwater, there has to be sufficient biodegradable organic matter to keep the micro-organisms alive (or active).

The biodegradable organic carbon component of the feedwater is either defined by the BOD (biochemical oxygen demand), which is a fraction of the COD (chemical oxygen demand), and the biomass content by the MLVSS. The rate at which the substrate enters the tank is thus given by the volumetric flow rate multiplied by the BOD or COD concentration, and the amount of micro-organisms by the volume of the tank multiplied by the MLVSS. This means that the food to micro-organism concentration is defined as:

$$F{:}M = \frac{sQ_1}{V_{\text{tank}}X'} \tag{6.13}$$

where s is the substrate concentration (the BOD or the COD), and X' the MLVSS. If all concentrations are expressed in g/m^3 (or mg/L), the tank volume in m^3, and the flow in m^3/d the ratio has units of d^{-1} – which means it isn't actually a ratio, since a ratio is necessarily unitless.

EXAMPLE: F:M DETERMINATION

If the feed BOD concentration for the previous example is 250 mg/L and the MLVSS makes up 75% of the MLSS, what is the F:M ratio?

The MLVSS is given by:

$$X' = 0.75 \times \text{MLSS} = 0.75 \times 3.5\,\text{g/L} = 2.625\,\text{g/L}$$

According to Equation 6.13:

$$F{:}M = \frac{sQ_1}{V_{\text{tank}}X'}$$

where $Q_1 = 2{,}000$ m^3/h, or 48,000 m^3/d.

So: $F{:}M = 0.25 \times 48{,}000/(15{,}000 \times 2.625) = 0.305$ d^{-1}.

EXERCISE 6.12
An aeration tank operating at an HRT of 12 hours maintains an MLSS of 6500 mg/l, of which 70% is volatile. What is the F:M ratio if the feed BOD concentration is 360 mg/L?

Chapter 7
Mass transfer and sedimentation

7.1 MASS TRANSFER

Mass transfer arises when there is a *driving force* capable of generating flow or else causing motion. In hydraulics the driving force is a pressure difference (or *pressure gradient*) which results in a flow of fluid. In electricity a voltage gradient results in a flow of electrical current. In mass transfer the driving force is a concentration gradient causing a flow of mass, and in heat transfer it is a temperature gradient and the flow is heat. The flow of all these quantities is subject to resistance imposed by bulk properties of the medium or by elements of the system. Thus, in all cases, these phenomena follow the general relationship:

$$\text{flow} = \text{driving force/resistance.} \tag{7.1}$$

Mass transfer is an important concept in water treatment since it controls many treatment aspects including aeration in an ASP, stripping of gases and volatile organic matter from water, and adsorption of organic matter or dissolved ions onto activated carbon or ion exchange materials (Section 4.6.5). It occurs wherever there are two or more components in a mixture and where those components are present at different concentrations within the mixture. If the mixture is a single phase then the mass transfer processes are relatively predictable. Heterogeneous systems, however, present phase boundaries which introduce complications.

There are two mass transfer mechanisms: *convection* and *diffusion*. Convective mass transfer occurs in mixed fluid systems, where mechanical or *forced convection* rapidly moves the molecules around the system at rates much higher than those achieved by diffusion alone. Density differences, caused by a temperature gradient, can also cause bulk fluid motion and so accelerate mass transfer over that from diffusion alone. The extent of influence of convection is evident from the ratio of momentum transfer (the kinematic viscosity $v = $ viscosity/density) to the *diffusion coefficient D*. This ratio is represented by the dimensionless *Schmidt number $Sc = v/D$* (see Table 2.3). For a component with a D of around 10^{-9} m^2/s in water ($v \sim 10^{-6}$ m^2/s), Sc is about 1000. In contrast, for gases, where diffusion is much more dominant, the value of Sc is around 0.1.

The most important mechanism of mass transfer in near-stagnant systems, where there is minimal convection, is that of *molecular diffusion*. Molecules vibrate in a random manner. High concentrations of a particular component mean that more of its molecules move in one direction (down the *concentration gradient*) than the reverse direction. The result is a gradual movement of molecules from high concentrations until a uniform concentration exists throughout the mixture. For a stagnant rectangular volume of the mixture of cross-section A and length l separating concentrations c_1 and c_2, the rate of mass transfer (in kg/s) by diffusion under steady-state conditions is given by:

$$\text{Rate} = DA(c_1 - c_2)/l \tag{7.2}$$

Table 7.1 Diffusion coefficient values.

System	D, $(m^2 \cdot s^{-1})$
Gases in gases at 273°K	$(0.5\text{–}2.5) \times 10^{-5}$
Oxygen in nitrogen	1.81×10^{-5}
Liquids in liquids	$(0.5 - 4.0) \times 10^{-9}$
CO_2 in water at 20°C	1.77×10^{-9}
Ammonia in water at 20°C	2.5×10^{-5}
NaCl in water at 18°C	1.26×10^{-9}
Ethanol in water at 10°C	0.83×10^{-9}
Oxygen in water at 25°C	2.5×10^{-9}
Solids in solids	$10\text{–}12 \times 10^{-34}$

where the units for the rate depend on those of the concentration, and the diffusion coefficient value D depends on the system (Table 7.1).

EXAMPLE: AMMONIA ABSORPTION

Ammonia is absorbed in water from a mixture with air at atmospheric pressure at 20°C. If the mass transfer resistance arises in a 1.2 mm thick gas film, what is the mass transfer rate per unit area if the partial pressure of ammonia is 0.08 atm? Assume that 1 mole of gas occupies 24 litres at the temperature of operation.

The fractional volume concentration is given by:

partial pressure/total pressure $= 0.08\,\text{atm}/1\,\text{atm} = 0.08$

Since the molar gas volume is 24 L/mol, the molar ammonia concentration is c_1:

$c_1 = 0.08/24 = 0.00333\,\text{mol/L} = 3.33\,\text{mol/m}^3$

According to Table 7.1, the diffusion coefficient for ammona in water is $2.5 \times 10^{-5}\,m^2 \cdot s^{-1}$ at 20°C. From Equation 7.2, assuming the interfacial concentration c_2 is zero, the rate of mass transfer per unit area is:

$N/A = Dc_1/l = (2.5 \times 10^{-5} \times 3.33)/(1.2 \times 10^{-3}) = 0.0694\,\text{mol} \cdot m^{-2} \cdot s^{-1}$

7.2 MASS TRANSFER AND PHASE BOUNDARIES

A number of two phase systems are encountered in water treatment, including gas–liquid (in ozonation and gas stripper columns), solid–liquid (in adsorption beds and in chemical precipitation) and gas–liquid–solid (in biotreatment processes such as trickling filters and the activated sludge process). Whilst mixing in the bulk of the separate phases may maintain a relatively homogeneous distribution of component concentrations within them, at the actual interface between the phases the relative velocity is zero and so convective transport is minimal. There therefore exists a region close to the interface where the relative velocity changes from zero (at the interface) to the bulk fluid velocity. Since there is also mass transfer of constituents across the interface, there must also be a change in their concentration across this region.

This zone is not completely stagnant, and so mass transfer is not related exclusively to diffusion. Since the zone width l is not known, a more general representation of Equation 7.2 is:

Rate of mass transfer, kg/s $= kA(c_1 - c_2)$ (7.3)

where k is a *mass transfer coefficient* (m/s) embodying all that is unknown about the velocity profile width and its effect on molecular diffusion. Mass transfer coefficients must thus be measured for each geometry and fluid flow characteristics of each multi-phase system.

> **EXERCISE 7.1 OXYGEN TRANSFER ACROSS A BOUNDARY**
> *If oxygen is transferred at a rate of 90 mg per minute across a 2 m² boundary where the concentration either side is 2 and 9.5 mg/L, what is the mass transfer coefficient?*

Because of the importance of gas–liquid contact devices both for reaction purposes (e.g. oxygen in an ASP) and for gas cleaning (e.g. acid gas scrubbing), much effort has been directed at understanding the mechanisms of mass transfer across a gas–liquid interface, though the concepts are equally applicable to other phase interfaces. In the *two-film theory*, it is assumed that two thin films exist at the phase interface (Figure 7.1) and that all mass transfer across the interface takes place within these two films. Outside the two films the bulk phases are considered to be homogenous (i.e. perfectly mixed).

The rate of mass transfer per unit area (N) across the gas, liquid and combined films is given by at any point i within the interface is:

Gas film $\quad N = k_G(p - p_i)$ $\hfill$ (7.4)

Liquid film $\quad N = k_L(c_i - c)$ $\hfill$ (7.5)

Overall $\quad N = K(p - c)$ $\hfill$ (7.6)

where k_G, k_L and K are the respective gas-film, liquid-film and overall mass transfer coefficients, and c and p refer to component concentration and partial pressure respectively. Equation 7.6 contains mixed units in p and c, which can be correlated under equilibrium conditions using Henry's Law (Section 4.6.4). Thus the equilibrium interfacial concentration c^* and partial pressure p^* are respectively defined by:

$$c^* = p/H \hfill (7.7)$$

and $\quad p^* = Hc$ $\hfill$ (7.8)

From these equilibrium concentrations and Equations 7.4 and 7.5:

$$N = K_G(p - p^*) \hfill (7.9)$$

and $\quad N = K_L(c^* - c)$ $\hfill$ (7.10)

where K_G and K_L are overall mass transfer coefficients respectively based on gas-phase and liquid-phase control. If there is no resistance to mass transfer in the interface between the two films, and equilibrium exists, then Henry's Law applies:

$$p_i = H_i c_i \hfill (4.32)$$

The overall coefficients then relate to the film coefficients according to:

$$\frac{1}{K_G} = \frac{p - p^*}{N} = \frac{p - p_i}{N} + \frac{p_i - p^*}{N} = \frac{1}{k_G} + \frac{H(c_i - c)}{N} \hfill (7.11)$$

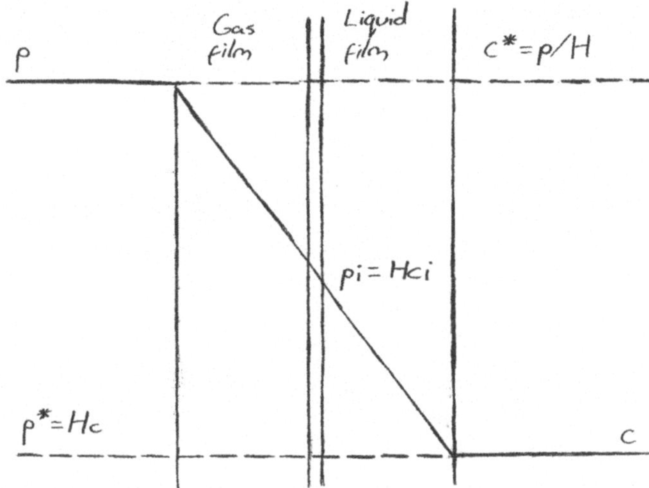

Figure 7.1 *Two-film* model of the gas–liquid interface.

or $\quad \dfrac{1}{K_G} = \dfrac{1}{k_G} + \dfrac{H}{k_L}$

and similarly, for the liquid film:

$$\frac{1}{K_L} = \frac{1}{k_L} + \frac{1}{Hk_G} \tag{7.12}$$

For highly water-soluble gases such as carbon dioxide, ammonia and hydrogen sulphide, H is relatively small and the second term in Equation 7.12 can be ignored, giving $K_G = k_G$. This case is referred to as *gas-film controlled*, and all of the concentration gradient exists in the gas film alone. For a relatively low-solubility gas like oxygen, H is large (Table 4.5) and the last term of Equation 7.13 is ignored, giving $K_L = k_L$. The gas transfer process is then liquid-film controlled.

7.3 ADSORPTION COEFFICIENT AND SHERWOOD NUMBER

If a volume of liquid is being contacted by a gas stream in a stirred vessel then the mean rate of mass transfer for the whole contactor, N_T, is given by:

$$N_T = k_L A \Delta c \tag{7.13}$$

where the area A is the total gas–liquid interfacial area and Δc the average concentration difference between the bulk of the liquid and the interface. The gas transfer rate per unit volume, N, is then given by:

$$N = k_L a \Delta c \tag{7.14}$$

The liquid-film coefficient is generally considered to be a constant value independent of stirrer speed and air rate. It is the specific interfacial area (a) in m^2/m^3 that changes due to the change in gas hold-up and the mean bubble size. Because 'a' cannot be easily measured in most practical applications, it is combined with k_L to form the *absorption coefficient, $k_L a$* (h^{-1}). The value of N is divided by Δc to obtain $k_L a$, where Δc can be obtained from:

$$\Delta c = c * E - c \tag{7.15}$$

where $c * E$ is the mean equilibrium concentration at the interface and c the mean liquid concentration. For oxygen c can be monitored using a DO probe.

For the gas phase, $c * E$ can be written as given by Henry's Law as p/H. If the gas phase is well mixed then the mean partial pressure p equates to p_{out} but for plug-flow reactors (Section 8.4.2) the log mean partial pressure difference (p_{lm}) must be used:

or $\quad p_{\text{lm}} = \dfrac{(p_{\text{in}} - p_{\text{out}})}{\ln(p_{\text{in}}/p_{\text{out}})} \tag{7.16}$

A value for the oxygen absorption coefficient ($k_L a$) enables design calculations to be carried out, by equating the oxygen supply rate to the steady-state BOD oxidation rate. The $k_L a$ values are obtained by measurement, or provided by equipment suppliers, for many gas–liquid contact geometries and methods of operation, such as mechanical surface aerators, diffused air systems, packed and sieve tray towers, and simple sprays.

EXAMPLE: AERATION

A 200 m^3 mixed aeration tank pressurised to 1.5 atm is fitted with bubble diffusers of absorption coefficient 2.2 h^{-1}. If the influent water has a DO concentration of 1.8 mg/L at 20°C, what is the rate of oxygen transfer?

Air contains ~20% oxygen by volume, and if the total pressure is 1.5 atm the partial pressure of oxygen in air, p_{O_2}, is $0.2 \times 1.5 = 0.3$ atm. According to Table 4.5, the Henry constant for O_2 at 20°C is 0.024 atm.L/mg. Thus:

$$c * E = p_{O_2}/H = 0.3/0.024 = 12.5 \text{ mg/L, or g/m}^3$$

The influent DO concentration c is 1.8 mg/L or g/m^3. So, for an absorption coefficient value $k_L a$ of 2.2 h^{-1}, the mass transfer rate according to Equation 7.14 is:

$$N = 2.2 \times (12.5 - 1.8) = 23.5 \text{ g/(h} \cdot \text{m}^3)$$

Since $V = 200$ m^3, the total transfer rate $N_T = 23.5 \times 200 = 4700$ g/h, or 4.7 kg/h.

EXERCISE 7.2

A 500 m^3 atmospheric plug-flow aeration tank is fitted with bubble diffusers having an absorption coefficient of 1.8 h^{-1}. If the influent water has a dissolved oxygen concentration of 2.4 mg/L at 20°C, and 60% of the oxygen in the air is dissolved, what is the oxygen transfer rate?

Mass transfer may also be expressed as the *Sherwood number*, Sh, another example of a dimensionless group (Table 2.3) and representing the ratio of mass transfer to diffusional velocity:

$$Sh = kd/D \qquad (7.17)$$

where d is the characteristic length of the system. The Sherwood number can be mathematically defined through the combination of other dimensionless groups. Such functions can be derived either empirically or from theory. For example, turbulent flow through a pipe is defined with reference to the Reynolds and Schmidt numbers:

$$Sh = 0.026\,Re^{0.8}\,Sc^{0.3} \qquad (7.18)$$

For bubbles in a stirred tank, mass transfer can be derived from the Archimedes and Schmidt numbers:

$$Sh = 0.13\,Ar^{0.3}\,Sc^{0.3} \qquad (7.19)$$

EXAMPLE: SHERWOOD NUMBER

What is the Sherwood number for dissolved CO_2 in water flowing at 2 m/s through a 25 mm pipe at a temperature of 20°C? [NB: see Tables 2.3, 3.5 and 7.1 for remaining data].

At 20°C the respective water density and viscosity values are approximately 1000 kg/m^3 and ~0.001 kg/(m · s) (Table 3.4) and thus the ratio (ρ/μ) is ~10^6.

From Table 7.1, $D = 1.77 \times 10^{-9}$ m^2/s for CO_2 at 20°C.

From Equation 7.18, $Sh = 0.026\,Re^{0.8}\,Sc^{0.3}$ where, according to Table 2.3:

$$Re = \rho v d/\mu = 10^3 \times 2 \times (25/1000)/10^{-3} = 50{,}000$$
$$Sc = \mu/\rho D = 10^{-3}/(10^3 \times 1.77 \times 10^{-9}) = 565$$

So $Sh = 0.026 \times 50{,}000^{0.8} \times 565^{0.3} = 0.026 \times 5743 \times 6.69 = 999$

EXERCISE 7.3

An aeration tank operates at 25°C. If the air density is 1.43 kg/m^3, calculate the Sherwood number for oxygen transfer, and hence the absorption coefficient, for a rising 2 mm bubble.

7.4 EXPERIMENTAL MEASUREMENT

Experimental measurement of mass transfer within a system usually proceeds by de-oxygenating the water and then measuring the rate of change of DO concentration with time (dc/dt). According to Equation 7.3, this is given by:

$$\frac{dc}{dt} = k_L a(c^* - c) \qquad (7.20)$$

where c^* is the bulk equilibrium concentration and c the concentration at time t. Integrating the above expression over time period t yields:

$$\ln\left(\frac{c^* - c}{c^* - c_0}\right) = k_L a t \qquad (7.21)$$

where c_0 is the original DO concentration (which should then be close to zero). Hence, a plot of the logarithmic term in Equation 7.21 against time yields $k_L a$, i.e. the absorption coefficient, in the slope.

7.5 SEDIMENTATION

Particles may be separated from water on the basis of their weight. Particles heavier (more dense) than water will sink, whilst those less dense will rise. If the water is quiescent or semi-quiescent then the particles can collect at the base or surface of the water, from where they may then be removed. Sedimentation (sometimes called *gravitation* or *settlement*) is the separation of particles heavier than water, whilst particles less dense than water are separated by *flotation*.

When a particle settles in water it does so at a constant velocity which arises from the balance of three forces. These comprise:

- The downward gravitation force arising from the density difference between the settling particle solids (ρ_s) and the water (ρ)
- The buoyancy force from the displacement of the water by the particle, which relates to the water density, and
- The drag force, which is a function of the particle cross-sectional area (and so its effective diameter d) and is dependent on the flow regime around the particle.

Because of the dependency of drag force on the flow regime the relationship between the settling velocity and size of a settling particle is itself dependent on particle size. Larger, denser particles settling more rapidly are associated with larger Reynolds numbers (Section 3.3) which pertain to more *turbulent* flow. However, for almost all practical considerations within water and wastewater treatment, *laminar* flow dominates during particle settlement and flotation. This being the case, the settling velocity v_s is given by *Stokes Law*:

$$v_s = \frac{g(\rho_s - \rho)d^2}{18\mu}$$

(7.22)

where g and μ are gravitational acceleration and liquid viscosity as before.

EXAMPLE: PARTICLE SETTLEMENT

What is the settling velocity of a 50 µm diameter silt particle of density 2500 kg/m³ in water at 20°C?

According to Equation 7.22, particle settling velocity is given by:

$$v_s = \frac{g(\rho_s - \rho)d^2}{18\mu}$$

At 20°C the water density and viscosity are ~1000 kg/m³ and ~0.001 kg/(m · s) respectively (Table 3.5). Converting all quantities to kg, m³ and s, the settling velocity is:

$$v_s = 9.81 \times (2500 - 1000)(50 \times 10^{-6})^2/(18 \times 10^{-3}) = 0.00204 \, \text{m/s} \; (\text{i.e.} 2.0 \, \text{mm/s})$$

EXERCISE 7.4
What is the approximate diameter of a particle of 1.23 g/ml density if it settles at 150 µm/s at 25°C?

Calculation of the sedimentation velocity v_s allows the flow rate though the sedimentation tank to be determined as the *surface overflow rate* v_o, defined as the ratio of the volume flow rate to the tank surface area:

$$v_o = Q_o/(lw)$$

(7.23)

where l and w are the length and width of a rectangular tank; v_o equates to the settling velocity v_s of the smallest particle the tank is able to retain. The required tank footprint (represented by lw) therefore decreases with increasing particle settleability.

EXERCISE 7.5
Calculate the length and the retention time of a rectangular sedimentation tank based on the removal of particles of 550 µm diameter and density 1020 kg/m³ entrained in water flowing at 30 m³/min through a tank 8 m wide and 4 m deep at 20°C.

Chapter 8

Reactor theory

8.1 INTRODUCTION

A reactor is a vessel in which a chemical or biochemical reaction or a physical separation process takes place. It may be identified by the type of reaction and/or vessel (hence an *oxidation column*, a *media filter bed*, or a *bioreactor tank*). The reactor may be operated either on a batch or continuous basis (Section 6.1).

It is normal in a batch process for the reactor constituents to be completely mixed or *homogeneous*: the concentrations of all components are the same at all points in the reactor. In a continuous process they may be either completely mixed (and the reactor is then referred to as a *continuous stirred tank reactor*, or *CSTR*) or else the contents are not mixed such that concentrations and reaction conditions change along the length of the reactor. If there is no mixing along the reactor length this is described as *plug flow*, and the reactor a *plug-flow reactor*, or *PFR* (Figure 8.1) Both batch and continuous processes are used in water and wastewater treatment, and both the type of reactor and the mode of operation of the continuous reactor (i.e. PFR or CSTR) influence the reaction rate and thus the reactant and product concentrations. As a result, the reactor size required for a given throughput depends on its configuration.

8.2 MASS BALANCE

A mass balance across the reactor shows the total mass flow entering a reactor to equal the total mass leaving (Section 6.1) plus the rate of accumulation (Equation 6.1):

$$\text{mass flow in} = \text{mass flow out} + \text{mass rate of accumulation.} \tag{6.1}$$

If, as is often the case in various water and wastewater reactor types (including disinfection, biological degradation, chemical oxidation and flocculation), the removal of a pollutant A in a reactor follows *first-order* kinetics (Section 5.2), then according to

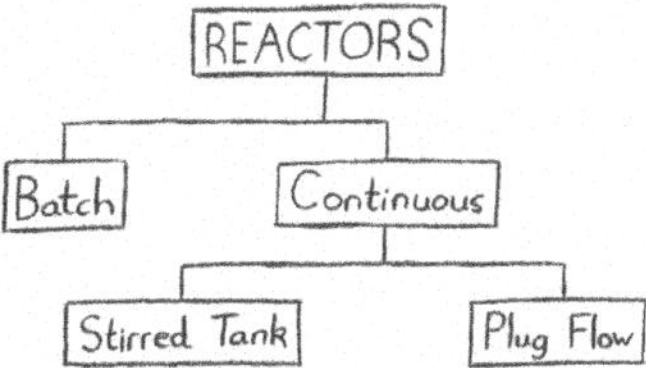

Figure 8.1 Reactor configurations.

Table 5.1 the rate r in kg m^{-3} s^{-1} at which it is removed is given by Equation 5.3 for a first-order reaction ($n = 1$):

$$r = -\frac{dc_A}{dt} = -kc_A \tag{5.3}$$

where c_A is the reactant concentration in kg m^{-3} and k the first-order rate constant in units of inverse time. If the mass m_A in kg of the reactant in the reactor of volume V is considered, then putting $c = m_A/V$ allows the above equation to be written as:

$$r = -\frac{1}{V}\frac{dm_A}{dt} = -\frac{dc_A}{dt} = -kc_A. \tag{8.1}$$

8.3 BATCH REACTORS

For a batch reactor, no product – the treated water or effluent – is available until the reaction is complete. To overcome the stop–start nature of product formation, multiple batch reactors are often operated in a cyclic manner so that there is always at least one reactor from which the product is being drawn. Batch reactors are often small in scale and are suitable for slower reactions requiring longer periods of time for the reaction to progress.

For a batch reactor there is no inlet or outlet flow and the reaction follows first-order kinetics, according to the equation listed in Table 5.1:

$$\ln(c_t/c_0) = -kt \tag{8.2}$$

where c_0 is the initial concentration (at time $= 0$), c_t the final concentration (at time t) and t the operation time.

In reality, operation of batch reactors involves long periods of downtime: unproductive periods when the reactor is being filled, emptied and cleaned. Furthermore, in biological processes, a lag time (Section 5.2.1) occurs at the start of the process during which there is either no or only very slow removal of pollutants.

EXAMPLE: DISINFECTION KINETICS

Coliform bacteria are inactivated at a rate constant of 0.18 min^{-1} according to first-order kinetics. If water containing 1000 cfu/100 ml ("colony forming units" per 100 ml) is to be disinfected so that the final coliform count is 1 cfu/100 ml, what contact time between chlorine and water is required for this?

From Equation 8.3, the initial and final coliform concentrations are given by:

$$\ln(c_t/c_0) = -kt$$

where $c_t/c_0 = 1/1000 = 10^{-3}$ and $k = 0.18$ min^{-1}.

So, $t = -\ln(10^{-3})/0.18 = 38$ min.

EXERCISE 8.1

A 500 m^3 batch sedimentation process has to achieve a 20 mg/L discharge limit for suspended solids against an influent concentration of 400 mg/L. If sedimentation proceeds at a rate of 3.5 × 10^{-4} s^{-1}, what is the time required to achieve the desired discharge limit? If the tank takes 36 min to empty and refill, how much water can be treated in a day?

8.4 CONTINUOUS REACTORS

Water and wastewater treatment reactors normally operate on a continuous, rather than on a batch, basis: there is a continuous flow of water into and out of the reactor. It is clearly advantageous to operate in this manner and, in such cases, time is not a variable as the system is under steady-state conditions (Section 6.2.1). Instead, it is the *residence* or *retention time* θ of the process which is of fundamental importance, this being the time that material remains in the reactor:

$$\theta = \text{tank volume liquid flow rate} = V/Q. \tag{8.3}$$

Hence, if the residence time of a continuous process were, for example, 2 hours then one reactor volume equivalent of feed would be processed every 2 hours.

Continuous reactors tend to require more control than batch reactors with higher associated capital costs. A continuous reactor is generally operated either as a CSTR or a PFR.

8.4.1 Stirred tank reactors

In the CSTR all the reactor vessel components are completely mixed during operation, such that the concentrations of the components in the reactor are the same as their concentrations at the reactor outlet (c_{out}). The CSTR is commonly used in water and wastewater treatment; typical examples in water treatment include neutralisation of acidic solutions and coagulation of colloids by chemical addition, and in sewage treatment the ASP (Figure 5a) is normally configured as a CSTR.

For a CSTR there is a continuous flow Q in and out of the reactor such that a mass balance can be conducted:

$$Qc_{in} = Qc_{out} - rV \qquad (8.4)$$

where r and V are respectively reaction rate (in concentration per unit time) and reactor volume.

Since the concentration, c_{out}, is uniform throughout the reactor, i.e. it is homogeneous, Equation 8.4 can be rearranged to:

$$(c_{in} - c_{out})/(-r) = V/Q = \theta \qquad (8.5)$$

For a first order reaction, the negative rate $-r$ equates to kc_{out} (Equation 8.1), hence:

$$\theta = (c_{in} - c_{out})/(kc_{out})$$

$$\text{or} \quad k\theta = (c_{in} - c_{out})/c_{out} = (c_{in}/c_{out}) - 1$$

$$\text{or} \quad c_{out}/c_{in} = 1/(1 + k\theta) \qquad (8.6)$$

8.4.2 Plug-flow reactors

In a plug-flow reactor (PFR) the feed enters the reactor and is converted into product whilst passing along its length. An ideal PFR has no longitudinal mixing so an individual *volume element* of fluid (i.e. a thin slice along the reactor length, Figure 8.2) does not interact with neighbouring elements in the reactor. As such, a PFR can be envisaged as a number of interconnected batch reactors each having the volume of the element. The component concentration, c, changes along the length of the reactor just as the concentration in a batch reactor changes with time, the distance along the reactor length being directly proportional to residence time.

In this case, there is a changing concentration along the reactor length. The mass balance is conducted by dividing up the reactor into the imaginary volume elements (dV) which are each considered as discrete but connected reactors over which a change in concentration (dc) takes place. The individual elements can then be summed by integration. Exactly the same mathematical process is used to produce Equation 8.2 from Equation 8.1, as well as the chemical kinetic expressions listed in Table 5.1 from differential equations such as Equation 5.3.

The mass balance is defined by:

$$Qc_{in} = Q(c_{in} + dc) - r\,dV \qquad (8.7)$$

which simplifies to the differential equation:

$$dc/dV = r/Q \qquad (8.8)$$

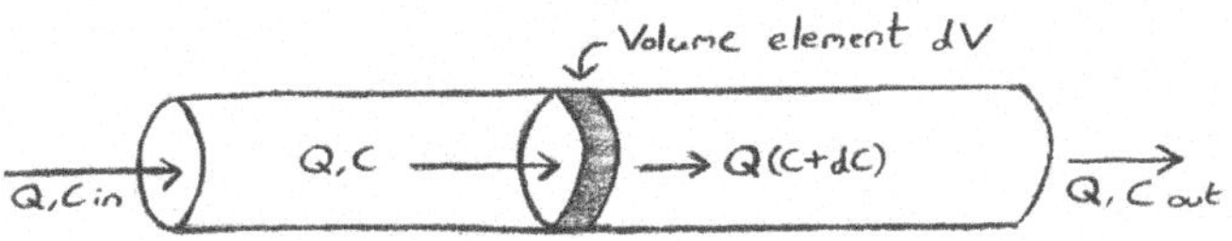

Figure 8.2 Plug-flow reactor.

Substituting r as before using Equation 8.1 yields:

$$dc/dV = -kc/Q \tag{8.9}$$

Integration of this expression, and substituting for V/Q as before produces:

$$\ln(c_{out}/c_{in}) = -k\theta$$

$$\text{or} \quad c_{out}/c_{in} = e^{-k\theta} \tag{8.10}$$

The equation thus takes the same form as that of the batch reactor, but with time being represented by the reactor residence time θ rather than the batch reaction time t.

8.5 COMPARISON OF REACTOR CONFIGURATIONS

Since water and wastewater treatment unit operations generally work on a continuous basis, it is useful to compare the impact of reactor configuration (CSTR vs. PFR) on tank size based on the same values for flow (Q), and feed (c_o) and treated water (c_{out}) concentrations. According to Equations 8.6 and 8.10:

$$\theta_{CSTR} = (c_{in} - c_{out})/(kc_{out}) = (c_{in}/c_{out} - 1)/k \tag{8.11}$$

$$\text{and} \quad \theta_{PFR} = -(1/k)\ln(c_{out}/c_{in}) = (1/k)\ln(c_{in}/c_{out}) \tag{8.12}$$

At the same flowrate, Q, and for the same reaction rate k, the ratio of residence times is:

$$\frac{\theta_{CSTR}}{\theta_{PFR}} = \frac{V_{CSTR}}{V_{PFR}} = \frac{((c_{in}/c_{out}) - 1)}{\ln(c_{in}/c_{out})} \tag{8.13}$$

Thus for the same physical or chemical/biochemical process (and thus the same value of k in Equations 8.11 and 8.12) a PFR is always more spatially efficient than a CSTR and, on this basis alone would always be preferred; the ratio of tank sizes depends on the c_{in}/c_{out} ratio, increasing from 1.02 at 5% conversion to 6.34 at 95% conversion.

However, in practice the choice between PFTR and CSTR is not based solely on reactor volume. Ideal plug-flow conditions (with no longitudinal mixing) are harder to achieve and maintain compared to ideal CSTR conditions. Most PFRs used in water treatment are of the media bed type in which water is contacted with some sort of granular media (such as adsorption processes, sand filtration and fixed-film biological processes). The flow conditions are often far removed from idealised tubular plug-flow reactor. In plug-flow ASPs, for example, the reactor cross-sectional area is usually sufficiently large for back-mixing, channelling and velocity fluctuations to occur. Also, in all municipal water and wastewater treatment plants, influent composition varies seasonally and/or diurnally (i.e. over the course of the day). Because of the rapid dilution of the influent stream in a CSTR these variations have less impact than they would on a PFR.

8.6 STIRRED TANK REACTORS IN SERIES

The performance of an ideal PFR can be approached by connecting multiple smaller CSTRs in series (Figure 8.3), which avoids some of the difficulties associated with PFR operation. The concentration entering each reactor is equal to the outlet (and reactor) concentration in the previous reactor; the concentration is reduced in a step-wise manner as opposed to the instant

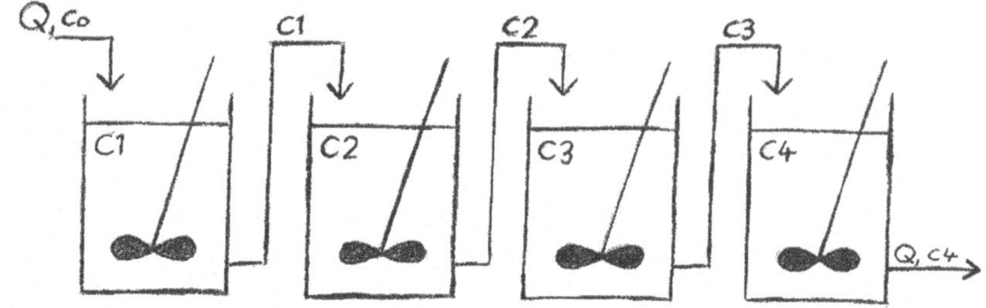

Figure 8.3 CSTR in series.

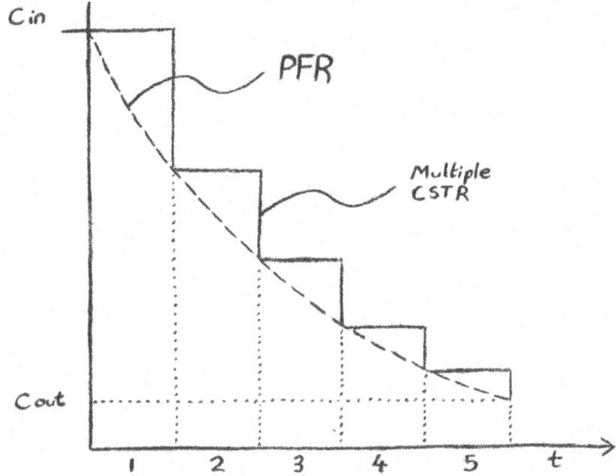

Figure 8.4 Comparison of five sequential CSTR reactors with a PFR.

reduction that would occur in a single large CSTR. Since the rate of reaction is proportional to the concentration in first-order reactions, for all but the final tank the rate of reaction is greater than arising in a single tank.

For a single CSTR, i, in the series fed from the outlet from the previous reactor, $i - 1$, according to Equation 8.6:

$$c(i - 1)/c_i = (1 + k\theta_i) \tag{8.14}$$

where all reactors are the same size and so have the same residence time θ_i. This expression can be extended to n reactors in series, where the total residence time θ is therefore $n\theta_i$, to give:

$$c_{\text{in}}/c_{\text{out}} = (1 + k\theta_i) = (1 + k\theta/n)^n \tag{8.15}$$

The impact of using CSTRs in series is shown in Figure 8.4. As the number of CSTRs is serially increased, the behaviour of the plant becomes more like that of a PFR which is, in effect, an infinite number of CSTRs in series. The step-wise reductions in the line that describes CSTRs in series would then become so small as to be the same as the curve that describes a single PFR.

EXERCISE 8.2
Assuming the same k value, calculate the ratio of the reactor volumes required to produce 99% removal using (a) a single CSTR, (b) six CSTRs in series, and (c) a single PFR.

8.7 FED BATCH REACTORS

Another variation which is found mainly in biological wastewater treatment is the *fed batch reactor*. This is, effectively, an automated batch reactor adapted for semi-continuous processing. The reactor is filled with reactant, processed and then the products are drawn out, hence the reactor is sometimes called a *fill and draw* reactor. Under these conditions, the reactor operates like the conventional batch reactor, analysed in Section 8.3. However, usually the reactor is not completely emptied at the end of the batch and some of the products from the last batch are deliberately left behind to dilute the influent such that it exhibits some of the characteristics of a CSTR.

Often wastewater treatment is conducted in a number of steps, usually in separate tanks for each step. However, they may also be carried out automatically in a single reactor as a *sequence*, in what is normally referred to as a *sequencing batch reactor* (SBR). A typical wastewater treatment sequence might consist of the following steps:

(1) Filling of reactor;
(2) Anoxic mixing for two hours to achieve *denitrification* (conversion of nitrate to nitrogen gas);
(3) Mixing with aeration for 16 hours for BOD reduction and nitrification (conversion of ammonia to nitrate);
(4) Settlement (or sedimentation, Section 7.5) for two hours to clarify the treated wastewater;
(5) *Decantation* (removal of the clarified supernatant) to discharge the treated wastewater;
(6) Sludge removal.

The principal drawback of the SBR is that in order to be able to handle a continuous flow a number of reactors (typically at least four) operating in parallel are required, each "out of phase" with the others. This significantly adds to the process complexity, demanding relatively sophisticated process control.

EXAMPLE: CSTR VS. PFR

A CSTR and PFR each have a reactor volume of 2 m³, and both treat the same wastewater at a flow rate of 5×10^{-3} m³/s with the same pollutant concentration. A first order reaction with a rate constant of 1.5×10^{-3} s⁻¹ applies to both reactors. Calculate the removal efficiency achieved in both reactors.

Removal efficiency is given by:

$$R = (c_{in} - c_{out})/c_{in} = 1 - c_{out}/c_{in}$$

Residence time is given by:

$$\theta = V/Q = 2/(5 \times 10^{-3}) = 400 \text{ s for both reactors}$$

For the CSTR (Equation 8.6):

$$c_{in}/c_{out} = 1 + k\theta, \text{ where } k = 1.5 \times 10^{-3} \text{s}^{-1}$$

Thus $\quad c_{in}/c_{out} = 1 + (400 \times 1.5 \times 10^{-3}) = 1.6$

So, $\quad R = 1 - 1/1.6 = 0.375$, or 38%

For the PFR (Equation 8.10):

$$c_{out}/c_{in} = e^{-k\theta}$$

Thus $\quad c_{out}/c_{in} = \exp(-1.5 \times 10^{-3} \times 400) = 0.549$

So, $\quad R = 1 - 0.549 = 0.451 = 45\%$.

EXERCISE 8.3

What is the ratio of tank volumes for a CSTR achieving 90%, 95% and 98% removal of influent biodegradable organic matter, assuming first-order reaction kinetics?

EXERCISE 8.4

A single reactor is being replaced by a train of 20 m³ reactors in series. The influent flow rate and concentration are 60 L/s and 150 mg/L respectively, and the reaction rate constant is 0.15 min⁻¹. How many reactors are needed in the new train to meet the discharge limit of 25 mg/L, and how much bigger is the volume of the original single reactor compared to the combined volume of the replacement reactors?

Chapter 9
Cost analysis

9.1 CAPEX AND OPEX

The determination of the cost of a project is normally through a consideration of the *capital* and *operating* expenditure (termed *CAPEX* and *OPEX* respectively). The CAPEX, or investment cost, refers to the one-off costs incurred at the start of the project. The OPEX is the cost associated with running the project. The *TOTEX* is then the total expenditure over the entire cost of the project. If the OPEX does not change with time, i.e. it has a fixed annual value, then the TOTEX is simply given by:

$$\text{TOTEX} = \text{CAPEX} + n \times (\text{OPEX per year}) \tag{9.1}$$

where n is the projected project life in years.

EXAMPLE: DOMESTIC SUPPLY OF FIZZY DRINKS

A household is considering buying a soda maker to supply itself with soda rather than buying bottles from the store. The household gets through three 2 L bottles a week on average, each costing $1.6. A soda maker costs $70 and lasts for 5 years. The consumables needed to make the soda are:

- Syrup packets, which cost $4 each and make 12 L of soda, and
- CO_2 canisters, which cost $30 and produce 75 L of soda

Is it worth switching to the soda maker?

The term of the project in this case is five years, based on the envisaged life of the soda maker. The number of bottles consumed is then given by:

$$N = \text{bottles per week} \times \text{weeks per year} \times \text{no. years}$$
$$= 3/\text{week} \times 365/7\,\text{weeks/y} \times 5\,\text{y} = 782.$$

The cost at $1.6 per bottle and volume (2 L/bottle) associated with this volume is:

$$\text{Cost} = 782 \times \text{Cost per bottle}\,(1.6) = 1251$$
$$\text{Vol} = 782 \times \text{Vol per bottle}\,(2\,\text{L}) = 1564\,\text{L}.$$

© IWA Publishing 2019. Watermaths: Process Fundamentals for the Design and Operation of Water and Wastewater Treatment Technologies
Author: Simon Judd
doi: 10.2166/9781789060393_0083

To generate this volume using the soda maker requires syrup and CO_2 canisters:

> No. syrup packs $= 1564/12 = 130.4$ packets
> No. CO_2 canisters $= 1564/75 = 20.8$ canisters.

If it is assumed that the remainder of the syrup packet and CO_2 canister maintains a residual value at the end of the five years then rounding up is not necessary. Multiplying the number required by the respective cost of each item ($4 for the sachets and $30 for the cannisters):

> Cost $= (130.4 \times 4) + (20.8 \times 30) = 1146.$

This has to be added to the cost of the soda maker ($70), making a total of $1216, or $35 less than the option of continuing to buy the pre-prepared soda. Given that this represents less than 3% of the total cost, the difference in overall cost is not significant. On the other hand, if the soda maker lasts for longer than five years its contribution to the total cost decreases with time.

9.2 DISCOUNTING

The previous example does not take account of the change in the value of money with time. While prices tend to increase with time due to inflation, the actual *value* of the money invested at the start of a project would be expected to increase according to the interest rate. Accordingly, if the original investment (S_0) were to be invested elsewhere and proportionally increase in value by D every year, its value (S_t) after t years would be:

$$S_t = (1 + D)^t \times S_0. \tag{9.2}$$

So, for example, a $100,000 investment made in 2010 would grow to a total of almost $180,000 in 2020 if subject to an annual 6% increase in value. By the same token, a sum of $180,000 spent in 2020 would equate to $100,000 in 2010 Figure 9.1. In the above equation, the parameter D is known as the *discount rate*. It takes account of the lost opportunity of investing the original capital sum in a financial scheme which would allow it to grow in value, and can be defined as the annual rate at which an invested sum S_0 must increase to achieve a target value S_t after time t. The difference in cost between two different investment options can be referred to as the *opportunity cost*, which is usually defined as the value of something that must be given up to acquire or achieve something else.

The invested sum S_0 is known as the *present value* or PV. Thus, an envisaged sum of $180,000 in ten years' time has a PV of $100,000 based on a discount rate of 6%. This means that any planned future spending on a project, such as routine maintenance or replacement of some critical component of a water or wastewater treatment unit operation, can be related to its present value at the start of the project. The true cost at the start of the project can then be evaluated, with delayed spending resulting in a financial saving due to the impact of the discount rate.

The purchasing power of any invested sum of money will be mitigated to some extent by the effect of inflation. For an inflation rate of i, the adjusted discount rate D_{real} is given by:

$$D_{real} = \frac{1 + D}{1 + i} - 1 \tag{9.3}$$

In practice the discount rate assumed may take account of inflation.

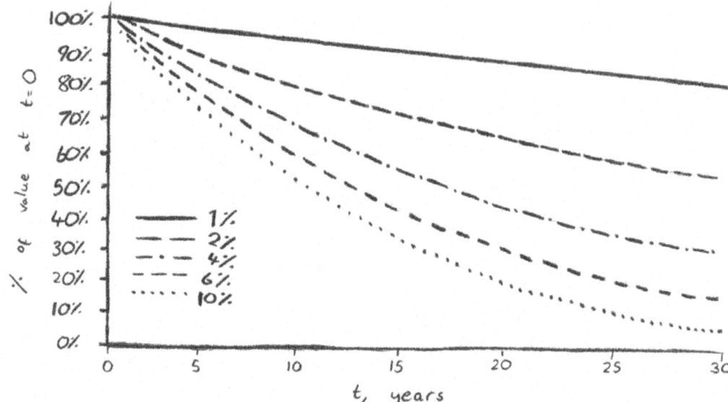

Figure 9.1 Change in value as a function of time at D values between 1 and 10%.

Equation 9.2 can be re-written to take account of the impact of discounting of future spending over the life n (in years) of the project:

$$S_0 = \sum_{t=0}^{t=n} \frac{S_t}{(1 + D_{\text{real}})^t} \tag{9.4}$$

In Equation 9.4 all contributions to the expenditure for each year of the project are adjusted for discounting by the factor $(1 + D_{\text{real}})^t$ and summed to produce the present value S_0. The costs at or before the start of the project are considered to be the CAPEX, whereas costs incurred thereafter make up the OPEX. Equation 9.4 then becomes:

$$PV = \sum_{t=0}^{t=n} \frac{\text{CAPEX}_{t=0} + \text{OPEX}_t}{(1 + D_{\text{real}})^t} \tag{9.5}$$

The calculation proceeds by determining the expenditure for each year between the start and the end of project when $t = n$, after which time its residual value has to be estimated. The residual value is often assumed to be negligible in comparison with all other expenditure, and may even be negative if, for example, decommissioning an installation incurs a cost.

If investing in the project leads to income, for example from selling the fizzy drinks from the previous example, then the difference between this income and the expenditure is known as the *net present value* (NPV). The total outgoings and income over the entire life of the plant then respectively contribute negatively and positively to the NPV.

EXAMPLE: DOMESTIC SUPPLY OF FIZZY DRINKS, COST ANALYSIS WITH DISCOUNTING

What is the present value of the two options given in the previous example?

The PV is determined by summing the discounted OPEX contributions in years 1–5 for both cases:

$$S_0 = \sum_{t=0}^{t=5} \frac{S_t}{(1 + D)^t}$$

Since the OPEX S_t does not change with time:

$$S_0 = S_t \sum_{t=0}^{t=5} \frac{1}{(1 + D)^t}$$

where, for the pre-made and home-made cases respectively:

$$S_t = 1.6 \times 3 \times 365/7 = \$250.3$$

and

$$S_t = (130.4 \times 4)/5 + (20.8 \times 30)/5 = \$229.1$$

The costs based on a discount factor of 6% can be calculated as follows, assuming the costs in the first year are discounted:

t:	0	1	2	3	4	5	*TOTAL*
$1/(1+D)^t$	1	0.943	0.89	0.839	0.792	0.747	
Pre-made							
CAPEX	0	–	–	–	–	–	
OPEX (bottles)		250.3	250.3	250.3	250.3	250.3	**1251**
PV	**0**	**236**	**223**	**210**	**198**	**187**	**1054**
Home-made							
CAPEX	70	–	–	–	–	–	
OPEX							
Syrup packets		104.3	104.3	104.3	104.3	104.3	**521.4**
CO$_2$ canisters		125.1	125.1	125.1	125.1	125.1	**625.7**
TOTAL OPEX		229.1	229.1	229.1	229.1	229.1	**1216**
PV	**70**	**216**	**204**	**192**	**181**	**171**	**1034**

> The PV is therefore reduced from \$1251 to \$1054 for the pre-made case and from \$1217 to £1034 for the home-made case, the latter including the capital cost of \$70. The difference between the two options is correspondingly reduced from \$35 to \$20.
>
> In common with all financial analyses of this type, the result is very sensitive to the assumed discount rate. For example, if the cost of the pre-made option is discounted at only 3% rather than 6% for the home-made option, perhaps because of differences in the rate of inflation between the individual components, then the total cost for this option increases to \$1146 and the cost saving for home-made soda increased to almost 10%.

In the case of a water or wastewater treatment plant, or any other infrastructure construction project, it is rare that the OPEX does not change from one year to the next over the life of the project. There is often scheduled refurbishment or component replacement, the timing of which impacts on the overall PV according to the discount factor and inflation assumed. The balance between CAPEX and OPEX, along with the assumed discount rate, then become critically important.

EXAMPLE: ROADBUILDING PROJECT

A civil engineering company is considering bidding for a contract to build and operate a toll road. The regulator has fixed the maximum charge to the toll road user over a 20-year period, after which time the company must re-bid for operating the road. Based on these figures for the maximum charge to the motorist, the company has calculated the following breakdown of the net income over five year tranches:

t, y	0	5	10	15	20
Revenue, m€	7	7.5	8.5	10	12

The company has also estimated:

- *An initial construction cost of €24 m*
- *A major refurbishment after 10 years costing €11*
- *A residual value of €2.5 m*
- *An overall average discount rate (D_{ave}) of 4%*

Is it worth bidding for the contract?

In this case it is instructive to consider the expenditure and income separately, discounting both for the five-year period to which they each refer. Whilst the undiscounted income significantly exceeds the expenditure by €12.5 m, or around 38% of the expenditure, the discounted totals indicate that the project would operate at a small loss of €0.4 m. Based on these figures, it would not be worth bidding for the contract.

t	Expenditure, €m			Income, €m			PV, €m	
y	CAPEX	OPEX	Residual			$1/(1 + D_{ave})^t$	Expend.	Income
0	24			7		100.0%	24	7.00
5				7.5		82.2%		6.17
10		11		8.5		67.6%	7.43	5.74
15				10		55.5%		5.55
20			−2.5	12		45.6%	−1.14	5.47
TOTAL, undiscounted		32.5		45		PV	30.3	29.9

The above example illustrates the important of discounting fixed future costs to a common point in time, this being the start of the project. All examples considered thus far are based on the assumption of a discount factor exceeding the inflation rate. As indicated in Equation 9.3, the impact of inflation is to reduce the discount rate. In cases where the inflation is higher than the discount rate then costs can be saved by investing at the start of the project. However, in most cases the costs incurred at some future point in the project lifespan are fixed and therefore have to be discounted back to the project start.

EXERCISE 9.1

A construction company has to make a decision as to whether to use incandescent, fluorescent or LED light bulbs for an office block under construction. Key data for the light bulbs are as follows:

Item	Incandescent	Fluorescent	LED
Cost (incl. deliv.) $	0.8	2	7.5
Power (W)	60	14	10
Life (h)	1200	8000	25,000

The following assumptions have been made:

- *A term of 8 years*
- *An average use of 8 hours a day*
- *An electrical energy cost of $0.12 per kWh*
- *An average labour cost of $40/h*
- *An average time of six minutes to change one light bulb*
- *An overall discount factor of 4%*

Assuming that there is no residual value in the light bulbs at the end of the period considered, which light bulb incurs the lowest cost over eight years?

There is very often a trade-off between CAPEX and OPEX in water and wastewater treatment. A more conservative design can increase the up-front investment costs but reduce the labour requirements or energy consumption, both of which contribute to the OPEX. For example, increasing the degree of automation can reduce the required staffing. Components offering a greater performance in terms of energy efficiency are likely to incur a higher cost.

9.3 OTHER FACTORS

It is often necessary to compare costs determined under different circumstances and convert them to a common reference point, usually the present day. The three most common parameters requiring such normalisation are time, currency and location.

The impact of time on the absolute cost can be accounted for by referring to the change in price of the item as a function of time. Strictly speaking, this should be carried out separately for all key items of investment, since the way in which an item changes with time depends on the nature of the item itself. Since this is rarely possible, recourse usually has to be made to the published annual *retail price index* (RPI) or *consumer price index* (CPI), which are measures of inflation. These indices represent averaged prices of selected goods and services normalised against some fixed point in time, and values at national level are available via various on-line source. The current price is thus simply the ratio of the current RPI to the RPI value for the year to which the reference data relates:

$$\text{Current price} = \text{Price at year Y} \times \text{RPI}_{current}/\text{RPI}_{yearY} \tag{9.6}$$

Prices can be converted to the required currency again using on-line data for currency conversion trends with time. For a cost given in a specific currency relating to a past year (Y), the cost must first be converted to the required currency using the currency conversion factor relating to Year Y before converting it to the current price using Equation 9.6.

Converting for location relies on specific information, namely *locations factors*, available only from specialist providers. Location factors account for changes in costs, principally construction costs, across different geographical regions of the world. These factors may be generated on a national or international basis.

Chapter 10

Solutions to exercises

CHAPTER 2 MATHEMATICS
EXERCISE 2.1

$$Re = \left(\frac{5.74}{10\sqrt{0.25/\lambda} - \kappa/(3.7d)} \right)^{1/0.9}$$

$$d = \frac{\kappa}{3.7\left(10\sqrt{0.25/\lambda} - 5.74/Re^{0.9} \right)}$$

EXERCISE 2.2

The velocity U can be expressed in terms of the Reynolds number:

$$U = Re \cdot \mu/(\rho d) = Re \times 0.001/(1{,}000 \times 0.1) = 10^{-5} Re$$

The Schmidt number can be calculated directly:

$$Sc = \mu/(\rho D) = 0.001/(1000 \times 1.9 \times 10^{-9}) = 526$$

Rearranging Equation 7.18 to isolate Re:

$$Re = \left(\frac{Sh}{0.26 \times Sc^{0.3}} \right)^{(1/0.8)} = \left(\frac{4{,}100}{0.026 \times 526^{0.3}} \right)^{(1/0.8)} = 300{,}000$$

So, $\quad U = 10^{-5} \times 300{,}000 = 3.00 \, \text{m/s}$.

© IWA Publishing 2019. Watermaths: Process Fundamentals for the Design and Operation of Water and Wastewater Treatment Technologies
Author: Simon Judd
doi: 10.2166/9781789060393_0089

EXERCISE 2.3

Allowing for the wall thickness:

- The internal height of the tank is $h - \delta$, where h is the tank height and δ is the tank wall thickness (assuming the tank base is the same thickness as the walls);
- The maximum internal tank diameter is $d_{max} - 2\delta$, where d_{max} is the minimum linear dimension within the available space (i.e. 8 m).

δ is 75 mm, i.e. 0.075 m. Therefore, the maximum volume is:

$$V_{max} = (h - \delta) \times \pi \times (d_{max} - 2\delta)^2 / 4$$

$$= (6 - 0.075) \times 3.141 \times (8 - 2 \times 0.075)^2 / 4$$

$$= 287 \, \text{m}^3.$$

The volume of a rectangular tank occupying the entire available floor area would be:

$$V_{max} = (h - \delta) \times (\text{width} - \delta) \times (\text{length} - \delta)$$

$$= (6 - 0.075) \times (8 - 0.075) \times (10 - 0.075)$$

$$= 466 \, \text{m}^3.$$

EXERCISE 2.4

In all cases the logarithm rules listed in Table 2.2 can be applied to give the following:

(a) $1.4b - a + \log c$
(b) $0.15 - 0.3ab + 0.7c$
(c) $0.3 + b + 7c - 2.3a$
(d) $\log a = m (\log d - \log b) - \log c$
(e) $t = (s/\log 2) \log (X/X_o)$

EXERCISE 2.5

This is an example of solution of simultaneous equations. Taking the log of the above equation produces:

$$\log p = \log k - n \log t$$

If the two sets of data are inserted into the equation, then two equations are produced:

$$-0.301 = \log k - 0.477 \, n,$$

and

$$-1.301 = \log k - 1.176 \, n$$

Subtracting the second of these from the first produces:

$$1 = 0.699 \, n, \text{ and so } n = 1.43$$

Resubstituting this value of n into either of the above two equations allows k to be calculated:

$$-0.301 = \log k - (1.431 \times 0.477), \text{ and so } \log k = 0.381 \text{ and thus } k = 2.4$$

EXERCISE 2.6

(1) Energy per unit mass $= \text{kg} \cdot \text{m}^2 \cdot \text{s}^{-2}/\text{kg} = \text{m}^2 \cdot \text{s}^{-2}$
     Power per unit mass flow $= \text{kg} \cdot \text{m}^2 \cdot \text{s}^{-3}/(\text{kg} \cdot \text{s}^{-1}) = \text{m}^2 \cdot \text{s}^{-2}$
     So, they have the same units.

(2)Energy per unit area per unit velocity $= \mathrm{kg} \cdot \mathrm{m}^2 \cdot \mathrm{s}^{-2}/(\mathrm{m}^2 \cdot \mathrm{m} \cdot \mathrm{s}^{-1}) = \mathrm{kg} \cdot \mathrm{m}^{-1} \cdot \mathrm{s}^{-1}$
These are the same units as those of dynamic viscosity (Section 3.2), i.e. derived SI units of $\mathrm{Pa} \cdot \mathrm{s}$

EXERCISE 2.7

Pipe cross-sectional area $A = (\pi/4)d^2$, where $d = 50\,\mathrm{mm} = 50 \times 10^{-3}\,\mathrm{m} = 0.05\,\mathrm{m}$.

So,$A = 3.141 \times 0.05^2/4 = 1.96 \times 10^{-3}\,\mathrm{m}^2$

Flow $= 30\,\mathrm{m}^3/\mathrm{h} = 30/3600\,\mathrm{m}^3/\mathrm{s} = 8.33 \times 10^{-3}\,\mathrm{m}^3/\mathrm{s}$

Velocity $=$ volume flow rate$/A = 8.33/1.96 = 4.25\,\mathrm{m/s}$.

The time to fill the tank, $\theta_w = V/Q$, the ratio of volume to flow rate, where the volume of the 12 m diameter and 5 m high cylindrical tank is given by (Table 2.1):

$V = \pi d^2\,L/4 = 3.141 \times 12^2 \times 5/4 = 565\,\mathrm{m}^3$

So, at a flow of 30 m^3/h, the time taken is:

$\theta_w = 565\,\mathrm{m}^3/30\,\mathrm{m}^3/\mathrm{h} = 18.8\,\mathrm{h}$.

EXERCISE 2.8

Density $=$ mass/volume, and thus mass flow $=$ density $\times$ volume flow

So,mass flow $= 1020\,\mathrm{kg/m}^3 \times 50\,\mathrm{m}^3/\mathrm{day} = 51000\,\mathrm{kg/d} = 51\,\mathrm{tonnes/d}$, or $2.13\,\mathrm{te/h}$

EXERCISE 2.9

$Pr = \mu C_p/k$, where μ has units of kg/(ms), C_p has units J/(kg $\cdot$ K), and k units of W/(m $\cdot$ K).

So, units are:

$$\frac{\mathrm{kg}}{\mathrm{ms}}\,\frac{\mathrm{J}}{\mathrm{kg\,K}}\,\frac{\mathrm{m\,K}}{\mathrm{W}}$$

K, m and kg all cancel, such that the expression reduces to J/Ws. This is then dimensionless, since a joule (J) is a watt-second.

CHAPTER 3 FLUID PHYSICS
EXERCISE 3.1

The Reynolds number is given by:

$$Re = \rho Ud/\mu$$

The velocity (U) is given by the ratio of the flow rate (Q) to the cross-sectional area (A), where:

$$A = (\pi/4)d^2$$
$$= (3.141/4) \times (50/1000)^2 = 0.00196 \, m^2.$$

So, $U = (25/3600)/0.00196 = 3.54 \, m/s$

Therefore

$$Re = 1000 \times 3.54 \times (50/1000)/0.001 = 1000 \times 3.54 \times 50$$
$$= 177,000$$

At 10°C the viscosity is 0.00131 (Table 3.5), or 31% higher than the assumed viscosity of 0.001 kg/(m · s) at 20°C. The Reynolds number would therefore decrease by 31%.

EXERCISE 3.2

The static head is being increased by 6 m and the pressure head reduced to zero. The energy Equation (3.8) thus becomes:

$$(6 + 6) + 0 + 0 = 1.5 + 0 + v^2/(2 \times 9.81)$$

So, $v^2 = 2 \times 9.81 \times (6 + 6 - 1.5) = 206,$

$$v = 14.4 \, m/s$$

and so the flow rate in the 75 mm pipe (cross-sectional area of $(\pi/4) \times 0.075^2 = 0.00442 \, m^2$) would be:

$$Q = 14.4 \times 0.00442 \times 3600 = 229 \, m^3/h.$$

To fill the unpressurised tank, at least 14.5 m head (1.45 bar) would be required to lift the water from the river, 2.5 m below ground level, to the tank 12 m above the ground.

EXERCISE 3.3

At the flow rate of 320 m³/h and diameter 0.3 m, the flow velocity in the pipe is:

$$(320/3600)/[(\pi/4)0.3^2] = 1.26 \, m/s$$

If a temperature of 20°C is assumed, and so density $\sim10^3$ kg/m³ and viscosity $\sim10^{-3}$ kg/(m⁻¹ s⁻¹), then the Reynolds number (Equation 3.7) can be calculated as:

$$Re = 10^3 \times 1.26 \times 0.3/10^{-3} = 3.77 \times 10^5$$

For ductile iron, having a pipe roughness of 0.05 mm (Table 3.1), the two terms in the denominator of Equation 3.11 are:

$$\kappa/(3.7D) = 0.05/(3.7 \times 300) = 4.50 \times 10^{-5}$$
$$5.74/Re^{0.9} = 5.74/(3.77 \times 10^5)^{0.9} = 5.50 \times 10^{-5}$$

The friction factor λ is thus given by:

$$\lambda = 0.25/\{\log[(4.5 + 5.5) \times 10^{-5}]\}^2 = 0.01.$$

From Equation 3.10, the hydraulic gradient s associated with flow through the pipe at this friction factor value is:

$$s = 1.25^2 \times 0.01/(0.3 \times 2 \times 9.81) = 0.00265 \text{ m head per m pipe length}$$

So, total head loss for a 2 km pipe is $0.00265 \times 2000 = 5.3$ m (or 0.53 bar).

EXERCISE 3.4

From Table 3.2, the sum of the k_{fitting} values for the six long radius bends ($k_{\text{fitting}} = 0.4$), and the other components (see Example in 3.5.1) is:

$$\Sigma k_{\text{fitting}} = (6 \times 0.4) + 0.5 + 0.8 = 3.7$$

The same Bernoulli balance as used in the example in Section 3.5.2 applies:

$$19.5 = \frac{v^2}{2\,g} + H_{f,\text{pipe}} + H_{f,\text{fittings}} = H_{f,\text{pipe}} + \frac{v^2}{2\,g}(3.7 + 1) = H_{f,\text{pipe}} + 4.7\frac{v^2}{2\,g}$$

where, for a 100 mm diameter pipe:

$$H_{f,\text{pipe}} = \lambda\frac{l}{d}\frac{v^2}{2\,g} = \lambda\frac{20}{0.1}\frac{v^2}{2g} = 200\lambda\frac{v^2}{2\,g}$$

Thus

$$19.5 = (200\lambda + 4.7)\frac{v^2}{2\,g};$$
$$v^2 = 2\,g \times 19.5/(200\lambda + 4.7) = 383/(200\lambda + 4.7)$$

So

$$v = \left(\frac{383}{200\lambda + 4.7}\right)^{0.5} \tag{A}$$

However, as before λ is a function of the Reynolds number (Equation 3.11), and thus also velocity:

$$\lambda = \frac{0.25}{\left[log\left(\dfrac{\kappa}{3.7d} + \dfrac{5.74}{Re^{0.9}}\right)\right]^2}, \text{ where } Re = \frac{\rho v d}{\mu} \tag{B}$$

where in this instance $\kappa = 0.05$ mm for uncoated steel (Table 3.1) and the pipe radius is 100 mm. Again, the value of 0.0236 can be selected for λ. Thus, from Equation A:

$$v = \left(\frac{383}{200 \times 0.0236 + 4.7}\right)^{0.5} = 6.38\frac{\text{m}}{\text{s}}, \text{ and so } Re = \frac{1000 \times 6.38 \times 0.075}{0.001} = 4.79 \times 10^5$$

Entering this into Equation B:

$$\lambda = \frac{0.25}{\left[log\left(\dfrac{0.05}{3.7 \times 100} + \dfrac{5.74}{479,000^{0.9}}\right)\right]^2} = \frac{0.25}{[log(1.35 \times 10^{-4} + 4.43 \times 10^{-5})]^2} = 0.0178$$

Re-entering this value into Equation A produces:

$$v = \left(\frac{383}{(200 \times 0.0178) + 4.7}\right)^{0.5} = 6.81 \text{ m/s}$$

This value differs by >6% from the previously determined value of 6.38. One further iteration produces a velocity of 6.83 m/s, which is <1% different to the previous value of 6.81 m/s. So, this value can be taken as the correct answer.

EXERCISE 3.5

Static head at inlet $= -25$ m
Static head at end $= +2.5$ m
So, total static head $= 2.55 - (-25) = 27.5$ m

Flow rate $Q = 43$ m^3/h $= 0.0119$ m^3/s,

For a 100 mm diameter pipe, i.e. $d = 0.1$ m, the cross-section area:

$$A = (\pi/4)0.1^2 = 0.00785 \text{ m}^2.$$

So, flow velocity:

$$v = 0.0119/0.00785 = 1.52 \text{ m/s},$$

and so, assuming values of 10^3 kg/m^3 and 10^{-3} kg/(m $\cdot$ s) for density and viscosity:

$$Re = \rho U d/\mu = 10^3 \times 1.52 \times 0.1/10^{-3} = 1.52 \times 10^5$$

Hydraulic gradient (Equation 3.10):

$$s = \lambda\, v^2/(2gd),$$

where, from Equation 3.11, the friction factor λ:

$$\lambda = 0.25/\{\log[\kappa/(3.7d)] + (5.74/Re^{0.9})\}^2,$$

and λ, according Table 3.1, is 0.05 mm for ductile iron.

So, $\quad \lambda = 0.25/\{\log[(0.05/(3.7 \times 100)) + (5.74/152000^{0.9})]\}^2$

$$= 0.25/\{\log[(1.35 + 1.24) \times 10^{-4})]\}^2 = 0.0194$$

Thus $\quad s = 0.0194 \times 1.52^2/(2 \times 9.81 \times 0.1) = 0.0228$ m/m

Since $L = 250$ m,:

$$H_{\text{pipe}} = 0.0228 \times 250 = 5.7 \text{ m head along length of pipe.}$$

k_{fitting} values, from Table 3.2, are:

long radius 90° bend	0.40
gate valve	0.12
sudden enlargement	1.00

So, total k values $= (6 \times 0.4) + 0.12 + 1.00 = 3.52$
So, $H_{\text{fittings}} = 3.52 \times 1.52^2/(2 \times 9.81) = 0.41$ m

Total pump head required $= 27.5 + 5.7 + 0.41 = 33.6$ m.

EXERCISE 3.6

Flow $= 300$ m^3/h $= 300/3600 = 0.0833$ m^3/s

Since the velocity $= 1.5$ m/s, the cross-sectional area of the flow is:

$$\text{Area} = \text{flow/velocity} = 0.0833/1.5 = 0.0556 \text{ m}^2$$

Channel width $w = 500$ mm $= 0.5$ m

Water depth in channel $d = $ area/width $= 0.0556/0.5 = 0.111$ m

Since hydraulic radius $R = $ area/wetted perimeter:

$$R = \text{area}/(2d + w) = 0.0556/(0.5 + 2 \times 0.111) = 0.0770 \text{ m}$$

According to Equation 3.15:

$$v = KR^{0.66}s^{0.5}$$

where, for smooth cement, $K = 60$ (from Table 3.3).

So, $1.5 = 60 \times 0.077^{0.66} \times s^{0.5}$
And so $1.5 = 11.0 \times s^{0.5}$
So, $s = (1.5/11.0)^2 = 0.0186$ m/m, or 19 mm/m

Note that s, the hydraulic gradient, is synonymous with the slope for an inclined channel.

EXERCISE 3.7

Flow $Q = 72 \text{ m}^3/\text{h}$, or $72/3600 = 0.02 \text{ m}^3/\text{s}$

All baffle channels are of the same width and thus have the same flow velocity:

$$v_1 = v_2 = \text{flow}/(\text{channel width} \times \text{channel depth})$$

So $v_1 = v_2 = 0.02/(0.25 \times 1.0) = 0.08$ m/s

From Equation 3.20, the loss through the baffles is given by

$$H_b \approx (1/2\,g) \times [N v_1{}^2 + (N - 1)v_2{}^2]$$

Since there are 15 baffles, i.e. $N = 15$:

$$H_b \approx [1/(2 \times 9.81)] \times (15 \times 0.08^2 + 14 \times 0.08^2) = 0.0095 = 9.5 \text{ mm}$$

According to Equation 3.21:

$$Q = K(L - 0.1\,N H_w) \times H_w{}^{1.5}$$

where $L = 0.5$ m, $Q = 0.02 \text{ m}^3/\text{s}$, $K = 1.84$ and H_w is the loss over the weir.
Since the weir covers the full length of the channel, $N = 0$ (Figure 12). Thus the above equation becomes:

$$Q = K L H_w{}^{1.5}$$

So, $0.02 = 1.84 \times 0.25 \times H_w{}^{1.5}$

Thus, the loss over the outlet weir is:

$$H_w = [0.02/(1.84 \times 0.25)]^{1/1.5} = 0.124 \text{ m} = 124 \text{ mm}$$

Since the inlet weir height is 1 m, this is the maximum depth of water in tank, and the cross-sectional area for the flow is:

$$A = 0.25 \times 1.0 = 0.25 \text{ m}^2$$

Hydraulic radius $R = $ area/wetted perimeter:

$$R = \text{Area}/(2d + w) = 0.25/(2 \times 1 + 0.25) = 0.111$$

From Equation 3.15, the hydraulic gradient s over the loss over the entire flow path length is given by:

$$v = KR^{0.66}s^{0.5}$$

where the velocity is the same as that for the baffle channels and, according to Table 3.3, for concrete $K = 40$. So:

$$0.08 = 40 \times 0.111^{0.66} \times s^{0.5} = 9.38\,s^{0.5}$$

and so $s = (0.08/9.38)^2 = 7.28 \times 10^{-5}$ m per m flow path.

For a total flow path length of 48 m, the head loss is:

$$H_f = 48 \times 7.28 \times 10^{-5} = 3.49 \times 10^{-3}\,\text{m} = 3.5\,\text{mm}$$

The total head loss equates to the difference in height between the weir levels:

$$H_{\text{total}} = 9.5 + 124 + 3.5 = 137\,\text{mm} = \text{inlet weir height} - \text{outlet weir height}$$

So if the inlet weir is 1000 mm above the base of the tank, the outlet weir height should be $1000 - 137 = 863$ mm.

EXERCISE 3.8

In this case, the following parameter values are affected:

μ = viscosity at 15 °C = 1.145×10^{-3} kg/(m · s) (from Table 3.5)
ρ = density at 15 °C = 999.13 kg/m^3 (from Table 3.5)
ε = porosity = 90% of 0.4 = 0.36
d = grain diameter = $(1 - 0.15) \times 0.55$ mm = 0.468 mm = 4.68×10^{-4} m
v_a = approach velocity = 1.2×7.5 m/h = 9/3600 = 2.5×10^{-3} m/s

So, the new head loss, again using Equation 3.23, would be:

$$H = 0.7 \times 5 \times (1.145 \times 10^{-3}/999.13) \times (2.5 \times 10^{-3}/9.81) \times [(1 - 0.36)^2/0.36^3] \times (6/4.68 \times 10^{-4})^2$$
$$H = 1.47\,\text{m}$$

Original head loss was 0.50 m, so the increase is almost 1 m of head, or nearly three times higher than the original value.

EXERCISE 3.9

The empirical equation for head loss is given by Equation 3.24:

$$s = H/l = 1.02^{(15-T)} v_a\, k_h$$

where $l = 1.5$ m. According to Table 3.6, 14/25 mesh sand has a head loss coefficient k_h value of 0.05.

Approach velocity v_a in m/h = flow/cross-sectional area = Q/A, where:

$Q = 30$ m^3/h
$A = (\pi/4)d^2 = (\pi/4) \times 1.8^2 = 2.54$ m^2

So, $v_a = 30/2.54 = 11.8$ m/h

Thus $H = 1.5 \times 1.02^{(15-12)} \times 11.8 \times 0.05 = 0.94$ m

CHAPTER 4 CHEMICAL STOICHIOMETRY AND EQUILIBRIA
EXERCISE 4.1

Mass concentration = Molar concentration × molecular (or ionic) weight
Molar weight of Na_2SO_4 is $2 \times 23 + 96$ (according to Table 4.1) = 142.
So, mass concentration = $142 \times 0.15 = 21.3$ g/L

EXERCISE 4.2

Equivalent weights, according to Table 4.2:

$$Ca^{2+} = 20, Mg^{2+} = 12, Na^+ = 23, HCO_3^- = 61, Cl^- = 35.5, SO_4^{2-} = 48$$

So, converting all cation concentrations to meq/L:

Ca = 118/20 = 5.9; Mg = 5/12 = 0.4.
Total = 5.9 + 0.4 = 6.3

And converting all anion concentrations to meq/L:

HCO_3 − 280/61 = 4.6; Cl = 57/35.5 = 1.6; SO_4 = 96/48 = 2.0.
Total = 4.6 + 1.6 + 2.0 = 8.2

Thus, equivalent concentration of sodium (Na) = 8.2–6.3 = 1.9 meq/L

Mass concentration of Na = $1.9 \times 23 = 44$ mg/L

EXERCISE 4.3

$pCa = -\log_{10}[Ca^{2+}]$, where $[Ca^{2+}]$ is the calcium concentration in moles/litre.
So,   if pCa = 2.5, then $[Ca^{2+}] = 10^{-2.5} = 0.00316$ M, or 3.16 mM

Since one mole of Ca weighs 40 mg/L (Table 4.1):

Ca concentration in mg/L = $40 \times 3.16 = 126$ mg/L

If 4 mM of Ca is added:
New Ca concentration = 3.16 + 4 = 7.16 mM, or 0.00716 M
So, new pCa = $-\log(0.00716) = 2.15$

EXERCISE 4.4

For this reaction, the reactants are NH_4^+ and HOCl, and the products NO_3^- and Cl^-

Oxidation number calculation

Assuming appropriate O.N. values for H and O:

Cl: From +1 in reactant to −1 in product, a change of –2
N: From −3 in reactant to +5 in product, a change of +8

Hence, for redox balance: 1 mole of NH_4^+ reacts with 4 moles of HOCl, and thus:

$$NH_4^+ + 4HOCl \Rightarrow 4NO_3^- + Cl^-$$

Charge balance

From the above equation, the charge balance is:

Reactants: $(1 \times +1) = +1$
Products: $(4 \times -1) + (1 \times -1) = -5$

Hence for charge balance, add 6 moles of negative charge (as OH^-) to reactants *or* 6 moles positive charge (as H^+) to product.

Material balance

Adding H^+ to the RHS:

$$NH_4^+ + 4HOCl \Rightarrow NO_3^- + 6H^+ + 4Cl^-$$

Since, N and Cl both balance only O and H need considering:

O: 1 mole deficit on product side
H: 2 mole deficit on product side

Thus, there is a deficit of one H_2O molecule on the RHS.

Hence the final balanced equation is:

$$NH_4^+ + 4HOCl \Rightarrow NO_3^- + 6H^+ + 4Cl^- + H_2O$$

EXERCISE 4.5

Analogous to Equations 4.35 and 4.29:

$$HFo \rightleftharpoons H^+ + Fo^-$$

Where HFo is formic acid and Fo^- is formate. For weak acids (Section 4.6.1):

$$pK_a = 3.75 = -\log \frac{[H^+][Fo^-]}{[HFo]} = pH + pFo - pHFo \qquad \text{A}$$

The total formic acid concentration is 50 mg/L. The molecular weight of formic ($HCHO_2$) is $12 + (2 \times 1) + (2 \times 16) = 46$ g/mol

Thus:

$$[HFo] + [Fo^-] = 50 \times 10^{-3}/46 = 0.00109 \qquad \text{B}$$

From Equation A at the original pH of 4:

$$3.75 = 4 + pFo - pHFo, \text{ and so } pFo - pHFo = -0.25$$

Interpreting the logs, given that "p" = "$-\log$":

$$[Fo^-]/[HFo] = 10^{0.25} = 1.78, \text{ or } [HFo] = [Fo^-]/1.78 = 0.562 \, [Fo^-] \qquad \text{C}$$

Substituting this into Equation B:

$$0.562[Fo^-] + [Fo^-] = 0.00109$$
$$[Fo^-] = 0.00109/(1 + 0.562) = 6.98 \times 10^{-4} \text{ M, or } 31 \text{ mg/L (for molar mass } Fo^- = 45).$$

Adding 50 mg/L HCl (molar mass $1 + 35.5$ g/mol) adds $50/36.5 = 1.37 \times 10^{-3}$ M H^+
Original $[H^+] = 10^{-4}$ M. So, new $pH = -\log (1.37 \times 10^{-3} + 10^{-4}) = 2.83$
So, from Equation A:

$$pK_a = 3.75 = 2.83 + pFo - pHFo$$

Or $[Fo^-]/[HFo] = 10^{(2.83-3.75)} = 0.120$, using the same manipulation as before
So $[HFo] = 8.32 \, [Fo^-]$, and $[HFo] + [Fo^-] = 0.00109$; the total concentration is unchanged
And so $[Fo^-] = 0.00109/(1 + 8.32) = 1.16 \times 10^{-4}$ M, or 5.2 mg/L (for molar mass $Fo^- = 45$).

EXERCISE 4.6

$K_{SP} = [Ca^{2+}][F^-]^2 = 3.2 \times 10^{-11}$ (Table 4.4)

Ca concentration $= 120$ mg/L, thus $[Ca^{2+}] = 120 \times 10^{-3}/40 = 0.003$ M (for a molar mass of 40 g, Table 4.1)

So $[F^-]^2 = 3.2 \times 10^{-11}/0.003 = 1.07 \times 10^{-8}$

and so $[F^-] = \sqrt{1.07 \times 10^{-8}} = 1.03 \times 10^{-4}$ M, or 1.96 mg/L for a molar mass of 19 g (Table 4.1).

EXERCISE 4.7

The molar mass of Pb is 207 g/mol. At 0.1 mg/L the molar concentration is therefore

$$M_{Pb} = 0.1/207 \text{ mM} = 0.483 \text{ }\mu\text{M, or } 4.83 \times 10^{-7} \text{ M}$$

According to Equation 4.32:

$$K_{sp} = [Pb^{2+}][OH^-]^2 = 4 \times 10^{-15} \text{mol}^3/\text{L}^3$$

$$pK_{sp} = pPb + 2pOH = 14.4$$

At pH 7, pOH $= 7$. Thus, if the solution is saturated in Pb:

$$14.4 = pPb + (2 \times 7)$$
$$pPb = 0.4$$
$$[Pb] = 0.398 \text{ M}$$

Since the actual concentration is only 4.83×10^{-7} M it is obviously unsaturated in the original solution at pH 7.

It is required to find the hydroxide concentration when $c_{Pb} = 0.1$ μg/L, i.e.

$$[Pb] = (0.1/207) \times 10^{-6} = 4.83 \times 10^{-10}$$

So, $pPb = 9.32$

Since $pPb + 2pOH = 14.4$

$2pOH = 14.4 - 9.32 = 5.08$

So, $pOH = 5.08/2 = 2.54$

Since $pH + pOH = 14$ (Equation 4.31)

$$pH = 14 - 2.54 = 11.45$$

So, for the Pb concentration to be reduced to 0.1 μg/L the pH must be increased to ~11.5, assuming it achieves equilibrium.

EXERCISE 4.8

Initial $[H^+] = 10^{-3.5} = 3.16 \times 10^{-4}$ M, which is less than the $[OH^-]$ added.

So, new pH $= 14 + \log([OH^-]_{added} - [H^+]_{existing}) = 14 + \log(0.0025 - 0.000316) = 11.3$

EXERCISE 4.9

Initial $[OH^-] = 10^{-(14-12.5)} = 0.0316$ M, which is less than $[H^+]$ added.

So, new pH $= -\log([H^+]_{added} - [OH^-]_{existing}) = -\log(0.04 - 0.0316) = 2.1$.

If only 0.02 M of acid was added the solution would remain alkaline:

Then, new pH $= 14 + \log([OH^-]_{existing} - [H^+]_{added}) = 14 + \log(0.0316 - 0.02) = 12.1$

EXERCISE 4.10

Henry constant for air (Table 4.5) $H_{air} = 0.046\ \mathrm{l \cdot atm/mg}$

At 1 atm, $c = 1/0.046 = 21.7$ mg/L
At 4 atm $c = 4/0.046 = 86.9$ mg/L
So, air released $= (86.9 - 21.7) = 65.2$ mg/L

EXERCISE 4.11

Free CO_2 concentration determined by Henry's Law.
From Table 4.5, $H_{CO2} = 2.72\ \mathrm{atm \cdot L/mol}$

At 1 atm the p_{CO2} is $(1 \times 0.03/100) = 3 \times 10^{-4}$ atm
So, $c_{CO2} = 3 \times 10^{-4}/2.72 = 1.1 \times 10^{-4} \mathrm{mol/L}$ (or M) $= [CO_2]$

From Equation 4.43:

$$[H^+][HCO_3^-]/[CO_2] = 4.14 \times 10^{-7}$$

From Equations 4.40-1: $[H^+] = [HCO_3^-]$ for CO_2 dissolution in pure water, and so:

$$[H^+]^2/1.1 \times 10^{-4} = 4.14 \times 10^{-7}$$

So, $\quad [H^+]^2 = 1.1 \times 10^{-4} \times 4.14 \times 10^{-7} = 4.55 \times 10^{-11}$

$\qquad [H^+] = \sqrt{(4.55 \times 10^{-11})} = 6.75 \times 10^{-6}$, and so pH $= -\log(6.75 \times 10^{-6}) = 5.2$

EXERCISE 4.12

$P = 120$, $M = 290$.
Since $P < \frac{1}{2}M$, from Table 4.7 the CO_3^{2-} concentration is $2P = 2 \times 120 = 240$ mg/L $CaCO_3$.

EXERCISE 4.13

In this case it is the value x, the dose of acid or base required to affect the pH change, which has to be calculated.

The initial alkalinity is $280/61 = 4.59$ mM and the pH is 8.1. $[CO_2]$ according to (Equation 4.45) is thus given by:

$$8.1 - 6.38 = 1.72 = \log(4.59 \times 10^{-3}/[CO_2])$$

So, $\quad 4.59/[CO_2] = 10^{1.72} = 52.5$, and so $[CO_2] = 0.0875 \times 10^{-3}$ M $= 0.0875$ mM.

According to Equation 4.46, adding x moles of HCl changes the pH thus:

$$\mathrm{pH} - 6.38 = \log([HCO_3^- - x]/[CO_2 + x])$$

So, based on the calculated HCO_3^- and CO_2 initial concentrations, to achieve the new pH of 7.0:

$$7.0 - 6.38 = \log([4.59 - x]/[0.0875 + x])$$

So, $\quad (4.59 - x)/(0.0875 + x) = 10^{(7-6.38)} = 4.16$

$\qquad 4.59 - x = 4.16(0.0875 + x) = 0.364 + 4.16x$
And so acid dose required $= x = (4.59 - 0.364)/5.16 = 0.82$ mM

For HCl, of molar mass $1 + 35.5 = 36.5$ g (Table 4.1):
So, $\quad$ dose $= 36.5x = 30\ \mathrm{mg/LHCl}$

EXERCISE 4.14

For phenol, according to Table 4.8, $K_F = 21$ and $1/n = 0.54$.

So, according to Equation 4.50 the adsorbed concentration of phenol is:

$$q_e = K_F c_e^{1/n} = 21 \times (15/1000)^{0.54} = 2.17\,\text{mg/g or } 2.17\,\text{g/kg}$$

The total bed capacity for a 1,000 kg column is then $1000 \times 2.17 = 2170$ g phenol
For 15 µg/L phenol, total volume treated is 2170 mg/0.015 mg/m^3 = 1.45×10^5 m^3
And, at 200 m^3/hr, the total run time is 1.45×10^5 m^3/200 m^3/hr = 725 hrs (~30 days)

For chloroform, according to Table 4.8, $K_F = 2.6$ and $1/n = 0.73$, and so:

$$q_e = K_F c_e^{1/n} = 2.6 \times (7.5/1000)^{0.73} = 0.0731\,\text{mg/g}$$

The total bed capacity for a 1,000 kg column is thus $1000 \times 0.0731 = 73.1$ g chloroform
For 7.5 µg/L chloroform, total volume treated is 73.1 g/0.0075 g/m^3 = 9750 m^3
And, at 200 m^3/hr, the total run time is 9750 m^3/200 m^3/hr = 49 hrs (~2 days).

CHAPTER 5 CHEMICAL KINETICS AND BIOKINETICS
EXERCISE 5.1

From Table 5.1, $k = -(1/t)\ln(c/c_0)$, where $c/c_0 = 1 - 0.36$ when $t = 10$ minutes
So, $k = (1/10) \times \ln(0.64) = 0.0446/\text{min}; t_{1/2} = \ln2/k = 15.5$ minutes
Resubstituting, $\ln(1 - 0.99) = -0.0446\,t$, so $t = 103$ minutes

EXERCISE 5.2

From Table 5.1:

$$c_t = c_0/(1 + ktc_0), \text{ and thus inverting both sides}: 1/c_t = 1/c_0 + kt$$

Substituting both sets of values into the inverted equation:

$$1/7.87 = 1/c_0 + 50\,k; \text{ and } 1/4.81 = 1/c_0 + 200\,k$$

Subtracting these two equations

$$1/4.81 - 1/7.87 = (200 - 50)k, \text{ thus } 0.0808 = 150\,k$$

So $k = 0.0808/150 = 5.39 \times 10^{-4}\,\text{L}/(\text{mmol}\cdot\text{s})$ or $0.539\,\text{L}/(\text{mol}\cdot\text{s})$ or $32.3\,\text{L}/(\text{mol}\cdot\text{min})$

EXERCISE 5.3

From Equation 5.5, $t_d = \ln2/\mu_m$, and thus $\mu_m = \ln2/t_d = \ln2/0.6 = 1.16\,\text{h}^{-1}$

From Equation 5.4, $\ln(x_t/x_0) = \mu t$, or $x_0 = x_t/e^{\mu t}$
After 24 h, $x_t = 3 \times 10^{18}$ cell/mL
So, $x_0 = 3 \times 10^{18}/e^{1.16 \times 24} = 3 \times 10^{18}/1.10 \times 10^{14} = 2.7 \times 10^6$ cells/mL.

EXERCISE 5.4

From Equation 5.9, the carrying capacity is given by:

$$c = \ln[(x_f - x_0)/x_0] = \ln[(3 \times 10^7 - 2.5 \times 10^6)/(2.5 \times 10^6)]$$

So, $c = \ln 11 = 2.40$

From Equation 5.8, for a time t of 2 hours and a specific growth rate μ of 0.005 min^{-1}:

$$x_t = x_f/(1 + e^{c-\mu t}) = 3 \times 10^7/(1 + \exp[2.4 - (0.005 \times 2 \times 60)]) = 3 \times 10^7/(1 + 6.05)$$

So, $x_t = 4.3 \times 10^6$ cells per ml

EXERCISE 5.5

From Equation 5.15:

$$M_{b,\max}/V = k_L a(c_{\text{DO}*} - c_{\text{DO,min}})/q_0\theta$$

The units of mg/L for DO concentration equate to kg/m^3

$$q_0 = 50\,\text{gO}_2/(\text{kg cells}\cdot\text{h}) = 0.05\,\text{g O}_2/(\text{kg cells}\cdot\text{h})$$

So $M_{b,\max}/V = 200 \times (5 - 0.5) \times 10^{-3}/(0.05 \times 8) = 2.25\,\text{kg cells}/(\text{m}^3\text{h})$

So, over a 24 h period, mass concentration generated $= 24 \times 2.25 = 54\,\text{kg m}^{-3}$
And for a 200 m^3 tank volume, total mass $= 54 \times 200 = 10800$ kg or 10.8 tonnes.

CHAPTER 6 MASS BALANCE
EXERCISE 6.1

Flow rate $Q_{\text{intermit.}}$ in m^3/h $= 25 \times 60 \times 60/1000 = 90$ m^3/h
Total time period during 1 h cycle water is pumped $= 2 \times 15$ minutes $= 30$ mins $= 0.5$ h
So, volume per h, $Q = 90 \times 0.5 = 45$ m^3/h

EXERCISE 6.2

Total cycle time:

$$t_{\text{cycle}} = 18 + 90/60 + 30/60 = 20 \text{minutes}$$

For flow of 12 L/s over 18 minutes, the total volume in litres delivered over the course of this filtration cycle is:

$$Q_{\text{filt}}t_{\text{filt}} = (12 \times 60) \times 18 = 12,960 \text{L}$$

Total volume of product water used for backflushing at 28 L/s for 90 s over the cycle is:

$$Q_{\text{backflush}}t_{\text{backflush}} = 90 \times 16 - 1,440 \text{L}$$

So, total production rate over 20 minute cycle is:

$$Q_{\text{net}} = (12,960 - 1,440)/(20 \times 60) = 9.6 \text{L/s (or } 34.6 \text{m}^3/\text{h})$$

So, the net production rate as a proportion of the filtration rate is 9.6/12 or 80%. This is usually referred to as the *recovery* or *conversion* of the membrane process.

EXERCISE 6.3

Calculating quantities of streams:

Feed: $Q_1 = 50$ m^3/d $= 50/24 = 2.0833$ te/h; $C_1 = 10$ kg/te
Filtrate: $C_2 = 50$ mg/L $= 0.05$ kg/te
Sludge: $C_3 = 4$wt% $= 40$ g/100 g $= 400$ kg/te

From Equation 6.6:

$$Q_2 = 2.083 \times (400 - 10)/(400 - 0.05) = 2.031 \text{ te/h (or } 0.0448 \text{ MLD)}$$

So, $Q_3 = Q_1 - Q_2 = 2.083 - 2.032 = 0.052$ te/h.

The mass flow of sludge solids ($C_3 Q_3$) is then given by:

$$M_3 = 0.052 \times 400 = 21 \text{ kg/h, or } 0.5 \text{ te/d}$$

EXERCISE 6.4

1 kg sludge originally contains:

    $1000 \times 0.71 = 710$ g water, and so
    $1000 - 710 = 290$ g solids.

60% of the original water is removed, such that the remaining water mass is:

    $(1 - 0.6) \times 710 = 284$ g water per kg of sludge.

Total mass of dried material $= 284$ g water $+ 290$ solids $= 574$ g
So the solids content in the dried sludge is $290/574 = 0.505$, or 50.5% by weight

EXERCISE 6.5

Solution density $= 1.12$ g/cm$^3 = 1120$ kg/m^3
So, mass concentration $= 45/1120 = 0.0402$ kg/kg

For solution mass flow $= Q$ kg/h:

   Mass flow of salt required $= 40$ kg/te $\times Q$ kg/h

   Mass flow of salt produced $= 5$te/12 h $= 5000/12 = 417$ kg/h

So, $40.2Q = 417$ and so Q $= 10.4$te/h $= 10.4 \times 1000/1120 = 9.26$ m^3/h

EXERCISE 6.6

Applying the mass balance for two feed streams and a single product stream as before and using units of g and m^3 (for convenience):

$$10 \times 80 + (Q_2 \times 10) = (Q_3 \times 50) \tag{i}$$

$$10 + Q_2 = Q_3 \tag{ii}$$

Substituting the expression for Q_3 into the Equation (i):

$$800 + 10Q_2 = (Q_2 + 10) \times 50 = 500 + 50Q_2$$

So, $Q_2 = (800 - 500)/(50 - 10) = 7.5$ m^3/h

EXERCISE 6.7

From Equation 6.7:

$$R_{\text{module}} = 0.5 = 1 - (1 - 0.095)^n$$

Rearranging and taking logs (Section 2.3):

$$\text{Log}(1 - 0.5) = n\log(0.905)$$
$$n = 6.94, \text{ or } 7 \text{ if rounded up to a whole number}$$

This means that the recovery per stage is:

$$R = 1 - (1 - 0.095)^7 = 0.503$$

For three stages:

$$R_{\text{overall}} = 1 - (1 - 0.503)^3 = 0.877 (\text{i.e. } 88\%)$$

EXERCISE 6.8

The average rate of flow into the tank is 45 m^3/h (according to Exercise 6.1)
The average rate of flow from the tank is 43.5 m^3/h
Rate of accumulation $=$ flow in $-$ flow out $= 45 - 43.5 = 1.5$ m^3/h
If the volume is 12 m^3 it will fill up in $12/1.5 = 8$ hours

EXERCISE 6.9

By analogy with the reaction given in the previous example between HCl and lime:

$$H_2SO_4 + Ca(OH)_2 \Rightarrow CaSO_4 + 2H_2O$$

The molecular weights of sulphuric acid and calcium sulphate, according to Table 4.1, are given by:

H_2SO_4: $(2 \times 1) + 32 + (4 \times 16) = 98\,g$

$CaSO_4$: $40 + 32 + (4 \times 16) = 136\,g$

One mole of $Ca(OH)_2$ (74 g molar weight) is required to neutralise one mole of H_2SO_4, the mass ratio is $74/98 = 0.755$. For an acid concentration (C) of 15% by weight and density (ρ) of $1100\,kg/m^3$, the mass flow of acid in kg/h is:

$Q(m^3/h) \times C(\textit{kg acid/kg water}) \times \rho(kg\ water/m^3 water) = 350 \times 0.15 \times 1100 = 57750$

So, required mass flow of $Ca(OH)_2 = 0.755 \times 57750 = 43{,}601\,kg/h$, or 43.6 te/h.
The calcium sulphate concentration in kg/m^3 (and so g/L) is given by the ratio of the mass flow of $CaSO_4$ in kg/h to the flow of water in m^3/h, including the water generated by the reaction which, i.e. two moles for each mole of $CaSO_4$ produced. So:

Mass flow $CaSO_4$ generated $= (136/74) \times 43601 = 80{,}132\,kg/h$

Mass flow water generated $= (2 \times 18/74) \times 43601 = 21{,}211\,kg/h$

The mass flow of feed water is given by the total mass flow minus the acid mass flow:

Mass flow of feed water $= (1100\,kg/m^3 \times 350\,m^3/h) - 57750\,kg/h = 327250\,kg/h$

So, the total volume flow of water generated, assuming a density of $1000\,kg/m^3$ for the product water, is given by:

Feed flow $+$ generated flow $= (327250 + 21211)/1000 = 348.5\,m^3/h$

So, product calcium sulphate concentration is given by:

Mass flow $CaSO_4$/volume flow water $= 80{,}132/348.5 = 230\,g/L$

If only 2 g/L is soluble then 228 g/L will be precipitated.

EXERCISE 6.10

Since 10% of the water is lost by evaporation the wastewater mass flow (Q_1) is 90% of the feed flow:

Q_1, $kg/d = 0.9 \times 300 \times 1000 = 270{,}000 = 270\,te/d$

For a mass flow of COD (M_1) of 600 kg/d, the feed concentration is:

$C_1 = 10^6 M_1/Q_1 = 600 \times 10^6/270{,}000 = 2222\,mg/kg$ or $mg/L = 2.222\,g/L$ or kg/te

If the permeate concentration is 70% of the feed, then:

$C_2 = 0.7 \times 2.222\,kg/te = 1.555\,kg/te$ (or 1555 mg/L)

And, since the permeate flow is 80% of the feed:

$Q_2 = 0.8 \times 270 = 216\,te/d$

And so $Q_3 = 270 - 216 = 54\,te/d$
Since $Q_1 C_1 = Q_2 C_2 + Q_3 C_3$
then

$$C_3 = (Q_1 C_1 - Q_2 C_2)/Q_3$$
$$= [(270 \times 2.222) - (216 \times 1.555)]/54 = 4.89\,kg/te\ (or\ 4890\,mg/L)$$

For the clarifier the feed stream is the retentate stream from the UF:

$Q_1 = 54\,te/d$

$C_{1,COD} = 4.89\,kg/te$ (prior to chemical dosing)
$C_2 = 20\,mg/L$, or $0.02\,kg/te$
$C_3 = 2\% = 20\,g/L = 20\,kg/te$

Chemical dosing with $FeCl_3$ adds to the feed suspended solids concentration conversion to ferric hydroxide, in a way analogous to Al hydrolysis (Equation 4.11):

$$FeCl_3 + 3H_2O \Rightarrow Fe(OH)_3 \downarrow + 3HCl$$

where the molecular weights are:

$FeCl_3$: $56 + (3 \times 35.5) = 162.5$
$Fe(OH)_3$: $56 + (3 \times 17) = 107$

So, 270 mg/L of $FeCl_3$ produces $(107/162.5) \times 270 = 178$ mg/L (or 0.178 kg/te) of $Fe(OH)_3$ precipitate and, since the ratio of COD to suspended solids is 2:1, each mg/L COD removed produces 0.5 mg/L precipitate. Dosing with ferric chloride removes 85% of the COD, so:

$$C_{COD\,precipitate} = C_{1.COD} \times \text{fraction removed} \times 0.5 = 4.89 \times 0.85 \times 0.5 = 2.08\,\text{kg/te}$$

Thus the total suspended solids concentration in the clarifier feed stream following Fe dosing is:

$$C_1 = 2.08 + 0.178 = 2.26\,\text{kg/te.}$$

From Equation 6.6, the clarified liquid flow is given by:

$$Q_2 = Q_1(C_3 - C_1)/(C_3 - C_2)$$

So, $\quad Q_2 = 54 \times (20 - 2.26)/(20 - 0.02) = 47.95\,\text{te/d}$
and so $Q_3 = Q_1 - Q_2 = 54 - 47.95 = 6.05\,\text{te/h.}$

EXERCISE 6.11

According to Equation 6.12:

$$\theta_s = \frac{V_{tank}X}{Q_2C_2 + Q_3C_3}$$

Rearranging this equation to isolate C_2, the treated water suspended solids concentration, gives:

$$C_2 = \frac{1}{Q_2}\left(\frac{V_{tank}X}{\theta_s} - Q_3C_3\right)$$

where $\theta_s = 9.5 \times 24 = 228$ h and $C_3 = 4.2\,\text{kg/m}^3$. So:

$$C_2 = \frac{1}{1950}\left(\frac{15{,}000 \times 3.5}{228} - 50 \times 4.2\right) = 0.0104\,\text{kg/m}^3 = 10.4\,\text{mg/L.}$$

EXERCISE 6.12

The HRT is given by the tank volume divided by the feed flow:

$$HRT = V_{tank}/Q_1 = 12\,\text{hours} = 0.5\,\text{d}$$

The MLVSS is given by:

$$X' = 0.7 \times MLSS = 0.7 \times 6{,}500\,\text{mg/L} = 4{,}550\,\text{mg/L}$$

According to Equation 6.13:

$$F:M = \frac{SQ_1}{V_{tank}X'} = \frac{Q_1}{V_{tank}} \times \frac{BOD}{X'} = (1/0.5) \times (360/4550) = 0.16\,\text{d}^{-1}$$

CHAPTER 7 MASS TRANSFER
EXERCISE 7.1

From Equation 7.3:

$$\text{Rate of mass transfer} = kA(c_1 - c_2)$$

Entering all data in units of g, m^3 and minutes:

$$0.09 \, \text{g/min} = k(\text{m/min}) \times 2 \, (m^2) \times (9.5 - 2) \, (\text{g/m}^3)$$
$$\text{So}, k = 6 \times 10^{-3} \, \text{m/min (or 6 mm/min)}$$

EXERCISE 7.2

For a plug-flow reactor the log mean partial pressure (Equation 7.16) must be calculated. In this case the inlet partial pressure $p_{\text{in}} = 0.2 \times 1 = 0.2$ atm and the outlet is 40% of this (since 60% is removed) and hence $p_{\text{out}} = 0.4 \times 0.2 = 0.08$ atm. So:

$$p_{lm} = (0.2 - 0.08)/[\ln (0.2/0.08)] = 0.131 \, \text{atm} = p_{O2}$$

So $c* = p_{O2}/H$ where $H = 0.024$ atm $\cdot$ L/mg, according to Table 17

$$c* = 0.131/0.024 = 5.46 \, \text{mg/L or g/m}^3$$

Since $c_o = 2$ mg/L or g/m^3 and $k_L a = 1.8$ h^{-1}, the mass transfer rate per unit volume according to Equation 7.13 is:

$$N_T = 1.8 \times (5.46 - 2.4) = 5.49 \, \text{g/(hm}^{-3})$$

Since $V = 500$ m^3, the total mass transfer rate is $5.49 \times 500 = 2745$ g/h.

EXERCISE 7.3

From Equation 7.19:

$$Sh = 0.13 \times Ar^{0.3} \times Sc^{0.3}$$

From Table 2.3:

$$Ar = d^3 \, g \, \rho \, \Delta\rho/\mu^2$$

and

$$Sc = \mu/\rho D$$

where
   $d = $ bubble diameter $= 2$ mm or 0.002 m
   $g = $ gravitational constant $= 9.81$ m s^{-2}
   $\mu = $ viscosity at 25°C $= 0.895 \times 10^{-3}$ kg/(ms) according to Table 3.5
   $\rho = $ density $= 997.07$ kg/m^3 according to Table 3.5

   So $\Delta\rho = $ liquid $-$ air density $= (997.07 - 1.43) = 995.64$
   Thus $Ar = 0.002^3 \times 9.81 \times 997.07 \times 995.64/(0.895 \times 10^{-3})^2 = 9.73 \times 10^4$

From Table 7.1, $D = 2.5 \times 10^{-9}$ m^2/s
So, $Sc = 0.8949 \times 10^{-3}/(997.07 \times 2.5 \times 10^{-9}) = 359$
And Sh $= 0.13 \times (9.73 \times 10^4)^{0.3} \times 359^{0.3} = 23.8$

From Equation 7.17:

$$k = Sh \times D/d = 23.8 \times 2.5 \times 10^{-9}/0.002 = 2.98 \times 10^{-5} \, \text{m/s (or 0.030 m/h)}$$

EXERCISE 7.4

According to Table 3.5, at 25°C:

Viscosity $= 0.895 \times 10^{-3}\,\text{kg/(ms)}$ and density $997.07\,\text{kg/m}^3$.

From Equation 7.22, the particle diameter is given by:

$$d^2 = 18\mu v_s/[g(\rho_s - \rho)]$$

Converting density to kg/m^3 and inserting values

$$d^2 = 18 \times 0.895 \times 10^{-3} \times 150 \times 10^{-6}/[9.81 \times (1230 - 997.03)] = 1.06 \times 10^{-9}$$
$$d = 3.25 \times 10^{-5}\,\text{m} = 32.5\,\mu\text{m}.$$

EXERCISE 7.5

From Equation 7.23, the surface overflow rate is given by:

$$v_o = Q_o/(lw)$$

where l and w are the tank length and width. v_o equates to the particle settling velocity:

$$v_s = \frac{g(\rho_s - \rho)d^2}{18\mu} = \frac{9.81(1020 - 1000)(550 \times 10 - 6)^2}{18 \times 0.001} = 0.00330\,\text{m/s}$$

assuming a temperature of 20°C. Thus, according to Equation 7.23, at a flow of 30 m^3/min and a tank width of 8 m:

$$v_o = 0.0033 = (30/60)/(8l)$$

So, $l = 19.0\,\text{m}$

The retention time is the ratio of the tank volume to the flow rate:

$$\theta = V/Q = lwh/Q = 19.0 \times 8 \times 4/30 = 20.3\,\text{minutes}$$

CHAPTER 8 REACTOR THEORY
EXERCISE 8.1

From Equation 8.3:

$$\ln(c_t/c_0) = -kt$$

So, $\quad t = (1/k)\ln(c_t/c_0) = -(1/(3.5 \times 10^{-4})) \times \ln(20/400) = 8559$ s or 143 min.

Total cycle time is given by:

$$t_{\text{cycle}} = \text{sedimentation time} + \text{downtime for emptying and refilling}$$

$$t_{\text{cycle}} = 143 + 36 \text{ minutes} = 179 \text{ minutes, or } 2.98 \text{ hours}$$

So, number of cycles per day $= 24/2.98 \sim 8$

Total volume treated per day is given by

$$Q = V_{\text{tank}} \times \text{cycles per day}$$

$$Q = 500 \times 8 = 4{,}000 \text{ m}^3/\text{d}.$$

EXERCISE 8.2

For 99% removal, $c_{\text{in}}/c_{\text{out}} = 100/1 = 100$

For a single CSTR, rearranging to Equation 8.6:

$$\theta_{\text{CSTR}} = (c_{\text{in}}/c_{\text{out}} - 1)/k = (100 - 1)/k = 99/k$$

For six CSTRs in series, according to Equation 8.15:

$$\theta_{\text{CSTR series}} = n((c_{\text{in}}/c_{\text{out}})^{1/n} - 1)/k = 6 \times (100^{1/6} - 1)/k = 6.93/k$$

For a PFR, according to Equation 8.10:

$$\Theta_{PFR} = -\ln(c_{\text{out}}/c_{\text{in}})/k = -\ln(1/100)/k = 4.61/k$$

EXERCISE 8.3

According to Equation 8.11 for a CSTR:

$$k\theta = (c_{\text{in}}/c_{\text{out}} - 1)$$

For 90% removal: $c_{\text{in}}/c_{\text{out}} = 100/(100 - 90) = 10$
So, $\quad k\theta = 10 - 1 = 9$

For 95% removal: $c_{\text{in}}/c_{\text{out}} = 100/(100 - 95) = 20$

So, $\quad k\theta = 20 - 1 = 19$

For 98% removal: $c_{\text{in}}/c_{\text{out}} = 100/(100 - 98) = 50$
So, $\quad k\theta = 20 - 1 = 49$

Thus, the ratio is 49 : 19 : 9, or 5.4 : 2.11 : 1.

EXERCISE 8.4

From Equation 8.3: $\theta = V/Q$, where the flow rate Q is 60 L/s, or 3.6 m^3/min (5.2 MLD)

The original reactor volume, according to Equation 8.6, is:

$$V_{\text{orig}} = Q\theta_{\text{orig}} = Q(c_{\text{in}}/c_{\text{out}} - 1)/k = 3.6 \times (150/25 - 1)/0.15 = 120 \text{ m}^3$$

For the 20 m^3 reactors, the residence time in minutes per reactor is:

$$\theta_i = V_i/Q = 20/3.6 = 5.56 \text{ minutes}$$

From Equation 8.15:

$$c_{\text{in}}/c_{\text{out}} = (1 + k\theta_i)^n$$

Taking logs and rearranging to isolate n:

$$n = \log(c_{\text{in}}/c_{\text{out}})/[\log(1 + k\theta_i)]$$
$$= \log(150/25)/[\log(1 + (0.15 \times 5.56)] = 2.95$$

So, the minimum number of tanks is 3, and these would provide a total volume of:

$$V_{\text{total,CSTR in series}} = 3 \times 20 = 60 \text{ m}^3$$

Thus, the new series is half the total volume of the original.

CHAPTER 9 COST ANALYSIS
EXERCISE 9.1

At a labour cost of \$40/h and a time of 6 mins (0.1 h) for changing a light bulb, the total cost of installing a light bulb is:

$$\text{Installation cost} = \text{purchase cost(\$)} + \text{labour cost} = \text{Purchase cost} + 0.1 \times \$40$$
$$= \text{purchase cost(\$)} + \$4$$

The energy consumption and cost is given by:

$$\text{Annual energy consumption, kWh} = \text{power (kW)} \times \text{(hrs per y)} \times \%\text{operating}$$

$$\text{Energy cost} = \text{total energy consumption} \times \text{specific energy cost (\$/kWh)}$$

where the %operating is 33.3%, based on 8 h/d, and the energy cost is \$0.12/kWh. The Annual energy cost is therefore given by:

$$\text{Annual energy cost} = \text{power} \times (365 \times 24) \times 0.333 \times 0.12 = 350 \times \text{power(kW)}$$
$$= 0.35 \times \text{power (W)}$$

The operational life in years is given by:

$$\text{Operational life (y)} = \text{Bulb life (h)}/(365 \times 24 \times 0.33) = \text{life (h)}/2920$$

The total number of bulbs needed is then given by the ratio of the project life to the operational life, the number being rounded up.

The labour costs per bulb are given by:

$$\text{Labour cost} = \text{(time in mins to change 1 bulb}/60)\times\text{labour hourly rate}$$
$$= (6/60) \times 40 = \$4 \text{ per bulb}$$

The calculated annual cost for each bulb are given below.

Item	*Incand.*	*Fluor.*	*LED*
Cost, \$ (incl. delivery)	0.8	2	7.5
Power, W	60	14	10
Bulb life, h	1200	8000	24,000
Operational life L, y	0.411	2.74	8.22
Number needed	**20**	**3**	**1**
Annual energy cost, \$	**21.0**	**4.91**	**3.50**
Maintenance cost per bulb, \$	**4.80**	**6.00**	**11.50**

The years during which a bulb must be replaced can be calculated from the ratio R_t of the number of elapsed years t to the bulb operational life. The number N of bulbs to be installed in any single year is then given by:

$$N = \text{ROUNDOWN}(R_t) - \text{ROUNDOWN}(R_{t-1})$$

where the function ROUNDOWN refers to the rounded-down integer of the number. This equation applies when the L is less than one year, i.e. for the incandescent bulb. In the case of the LED bulb the life is longer than the project life, such that only one bulb is needed. The fluorescence bulb has a life of 2.74 years, which means that it must be changed during the third and sixth year of the project.

In all cases the bulb must be installed at the start of the project, with the cost incurred by this representing the CAPEX. Any subsequent replacement of the light bulbs contributes to the OPEX and must be discounted with time. A summary of the discounted costs is given below, based on the assumed overall discount factor of 5%. The calculations include a replacement of the incandescent bulbs in the final year of the project.

Year:	0	1	2	3	4	5	6	7	8	Total
$1/(1 + D)^t$	100%	95%	91%	86%	82%	78%	75%	71%	68%	
Incandescent bulb										
N	1	1	2	3	2	3	2	3	3	**20**
Maintenance costs,$	4.8	4.8	9.6	14.4	9.6	14.4	9.6	14.4	14.4	**96**
Energy cost, $	21.0	21.0	21.0	21.0	21.0	21.0	21.0	21.0	21.0	**189**
PV, $	**25.8**	**24.6**	**27.8**	**30.6**	**25.2**	**27.8**	**22.9**	**25.2**	**24.0**	**234**
Fluorescent bulb										
N	1			1			1			**3**
Maintenance costs,$	6.0			6.0			6.0			**18**
Energy cost, $	4.91	4.91	4.91	4.91	4.91	4.91	4.91	4.91	4.91	**44.2**
PV, $	**10.9**	**4.7**	**4.4**	**9.4**	**4.0**	**3.8**	**8.1**	**3.5**	**3.3**	**52.3**
LED bulb										
N	1									**1**
Maintenance costs,$	11.5									**11.5**
Energy cost, $	3.50	3.50	3.50	3.50	3.50	3.50	3.50	3.50	3.50	**31.5**
PV, $	**15.0**	**3.3**	**3.2**	**3.0**	**2.9**	**2.7**	**2.6**	**2.5**	**2.4**	**37.7**

According to the assumptions made, the LED bulb presents the lowest present value cost. The numbers would be moderated to some extent by accounting for inflation, but given that the main contribution to the OPEX is energy demand and the energy source is the same for all three bulbs, the impact would also be proportionally the same.

ASSIGNMENT SOLUTION HINTS AND ANSWERS

(1) See Sections 3.5.1–3.5.2 and accompanying examples and exercises.
 $\lambda = 0.0153$; $H_f = 2.34$ m along pipe and 1.22 m through fittings, 3.56 m in total.

(2) See Section 2.2 for geometry expressions. Biotank is 4500 m^3 in volume, 21.2 m wide and 42.4 m long. From Equation 8.7, BOD_{out} is 14 mg/L

(3) Using Equation 7.14 to find gas transfer rate per unit volume, $N = 3.42$ g/(h·m^3). This equates to a required volumetric aeration rate of 100 Nm^3/h.

(4) See Section 4.6.5.2 and Exercise 4.13. From Equation 4.45 the initial CO_2 concentration is 5.29×10^{-4} M based on the initial bicarbonate concentration of 0.0044 M and the pH. Using Equation 4.46 the acid dose "x" required to reach a pH of 6 is 2.95 mM, or 3.47 tonnes H_2SO_4/day. However, see (5) below.

(5) Solids derive from the coagulant (Equation 4.11). Stoichiometric calculations reveal 57.8 mg/L of $Al(OH)_3$ is generated from a dose of 20 mg/L of Al. Also, because Al hydrolysis generates acid (or removes hydroxide, see Equation 4.11), 2.22 mM of acid is generated by dosing with 20 mg/L of Al. So, only a further 0.73 mM of acid is actually needed to complete the pH adjustment to 6 for (4) above.

(6) Use Equation 3.23 to determine v_a and l based on the feed flow rate to the filter (1/20th of the total flow from the bioreactor, which is 100−1.75% of the original feed flow). Flow = velocity × cross-sectional area A, and $A = \pi d^2/4$. The filter length and diameter values can then be determined as 1.53 and 2.55 m respectively.

(7) See Exercise 5.1. Contact time is 39.9 minutes demanding a tank volume of 326 m^3 for the full flow of 8.19 m^3/min.

(8) Assuming the hypochlorite is reduced to chloride (as in Exercise 4.4) and the bisulphite is oxidised to sulphate, balancing of the equation for the redox reaction (Section 4.5.1) reveals that bisulphite and hypochlorite react in a 1:1 ratio. For a 10% stoichiometric excess, 4.83 mg/L $NaHSO_3$ is needed to remove the residual.

(9) From Equation 6.7, the retentate flow is a factor of $(1 - \alpha)^n$ of the feed flow, where α is the conversion per element and n the number of elements per module. Use this to determine the maximum number of elements per module (5). The feed capacity of the element demands that there must be nine modules in the first stage. The complete array is (9 × 5) for Stage 1, (6 × 5) for Stage 2, and (4 × 5) for Stage 3. The resulting overall conversion is 71.4%, giving a permeate flow of 351 m^3/h.

(10) At a pH < 4.5 all alkalinity is converted to CO_2 (Section 4.6.5.1) and the feed CO_2 concentration is then 4.93 mM, as determined by summing the bicarbonate and CO_2 concentrations from Pt 4. Conducting a mass balance (Equations 6.4–6.5), based on the feed flow of 351 m^3/h from the RO permeate, reveals the outlet CO_2 concentration to be 5.9 mg/L.

IWA Publishing's authorised EU representative for General Product
Safety Regulations is Diane D'Arras, 15 rue Duret, 75116 Paris,
France, e-mail: safety@iwap.co.uk.

Printed and bound by CPI Group (UK) Ltd, Croydon, CR0 4YY

12/05/2026

02108886-0015